Space Age

Ólafur Gunnsteinsson

Strategic Book Publishing and Rights Co.

Strategic Book Publishing and Rights Co.
12620 FM 1960, Suite A4-507
Houston, TX 77065
www.sbpra.com

ISBN: 978-1-60976-642-9

Book Design by Julius Kiskis

21 20 19 18 17 16 15 14 13 1 2 3 4 5

Entelect

Reykjavík, 2001

Stability. Economic solutions. Warfare has never been considered a solution. A third cultural fusion. Land. Water. To save the cosmic conscience. Because of its important pact.
The eyes of the world.
Gate of paradise.
Key to the universe.
Solace.
Understand numerals. Vision the future.
-Not on an empty stomach, though.
Deadlocked in his own image.
The real theory of movement. Images: form and color shape pictorial replicas or re-evaluations or practical copies in either extremes.
The senses, as expressions of one's own existence; image, actions, small scope.
Odour: through space. Particles drawn together. How light, sound . . . appear as sense data/perception: duration in time/space
Timing: prolonging of intervals
Theories of determinations
No empirical dogmas. Rather crystal facts.
How things happen. Stages, phases, layers. Compound mass unit.
Exceeding the speed of mind would be a small kind of a revolution. (To say the least).

Mind vs. mind
Timing

Summary from an elevator.
Elementary forces, such as political and religious movements.
Negotiation

Cooperation	international
Equality	strict and proper
Stability	political

-Tolerance, patience, mutual respect
-Legalistic
Cultural differences. Unified ground from which to harvest.
:Increase cooperation
resources
borders
Economical principles.

Sorrow/annoyance	Joy
Submittance/humiliation	arrogance
Fright	anger

Patrimony of the blind the hermetic order.
The magic of will. A structure of superpower, random, magnetic, paranormal experience living through (other bodies).
Awareness as a monitor to unlock any kind of messages.
As an instrument. As attraction of colors.
Thou shalt not use force.
ind vs. machine
man in between.
So that no other innocent life will be sacrificed on the altar of stupidity, there is no other principle to hold than the aforementioned.
Cooperation.
To crack down on suicide bombers.
Sequence of orders: leave Palestinian territories.
Immediate: Cease-fire or leave territories.
At the table:

future lives?
immediate
leave territories
Future territories
immediate
cease-fire
Ethnic disarmament
disillusioned army.
Nation disqualified in advance
:don't pretend anymore. Heavily medicated democracy.
Basic stuff. Home and families. A life.
Center focal point: pitfalls
degrees of perception
illusion of both parties
disarmament impossible.
Replacement of territories unacceptable.
Religious, ethnic, cultural.
Immigration.
Hostility of neighbouring nations.
Overpopulated areas.
Minerals and resources.
Economical (poverty-wealth).
Historical background.
Different categories according to human rights.
different values
qualitative misconception.
No compromise at eye sight.
Mystical tendencies in both directions.
General claim for a capital city.
Cultural fusion impossible at the moment.
Wrong historical decisions have led to catastrophes.
The deep silence of the media

4

A craving for a new world vision
The Koran. (The cow, 2:22)
. . . Guard yourselves against the fire whose fuel
is man and stones, prepared for the unbelievers.
The inspinated nightingale.
Where the Zaccham flower grows.

2002

Witnessing the disarmament in progress.
(*Hospes sumus et aula nostra est terra*).
At a distance through an empty mirror.
Major trends
Different motivations. Cultural fusion. Esoteric, heretic. Phenomenal meditation. Different categorical perception. Classical, baroque. Breakthrough, feedback. Real theatre at its extreme. Spartan mimics. Alteration of limits. Zero tolerance. Oppressive network. Large-scale corporations. A=(=)B= ?
They pointed: Alexander to the end of the world
Napoleon to perdition
Mass control depriorization. Distribution of power.
The hierarchy of establishments
The medium is the dragon´s eye.
Few powers:
Executive power, legislative, judicial, capital, hidden, religious, authoritative, media, corporative, military power, mass power, real power, spiritual power.
Constants, principles. Fundaments. Substances.
Propositions. Attributes.
Tartaria, where the hidden word awaits.
The home of the wild horses, land without boundaries.
Over Saxon hills, beyond the lighthouse

(and still we were forced to sing: damn that stone . . .)
The cup was filled above the rim.
The void and now
Essence out of mind
The four freedoms:
Goods, corporations,
capital and people.
(Arts, sciences, religion, politics)
I. Berlin: Freedom from, freedom to
(feudal freedom)
Decisions making.
Immediacy: syntactic truths.
Volition of the past.
Time: past perception and immediate volition
Morph, form. Atomic. Metalanguage
L.Wittgenstein, Über geviBheit, I 17: *durchsichtige/ undurchsichtige farben. Farbengleichheit . . . wie durch ein farbloses gesehen werden. II 7:weiB aufhebt dunkelheit? III 3: eine art Farbenmatematik.*

2003

Marketplace arena
burns out cold
Reality in a more ideal size: thought and avoidance of time.
Search for low profile companies, fear the worst.
Why not take all the cake?
Dominion of the four.
Time integrating with the mind.
Incorporate unions, trademark unknown.
Given the warranty of legal aid system, complete overview.
Ground zero

Silence detector without bounds.
Surpass spiritual hindrances. Someone who knows.
The public disharmonizes every individual act.
Alien mongrels from wherever: the hypnotist is in the house
Surpass yr memory
It all comes down to this: space and void
shipment due to day,
various aircraft, missiles, info.
Filling up a void: the category of the lost and lonely
Yr magnetism surpasses boundaries
Mind and its limits.
Surpass yr mind.
Lose potency.
Diplomatic minds can turn vicious on you every moment now.
Created a gap between autonomies.
The missiles they've used were filled up with germs.
A large group has been hospitalized,
we have seen it happen it will happen again
Constellation of supernovas.
The galaxy is yrs.
A global unit.
Where dragons sleep.
Napoleon
still points to perdition.
Bukaphelos come home:
will you ever?
New market theory.
Absolute power of freedom.
Timing at all cost.
The spirit must be kept clean:
has to be pointed out
Form:

Density
Altitude
Position: state of mind
To impose on color as on form:
black to red to white
Kandinski: Yellow advancing. Qualities; eccentric,
active, warmth.
Blue, black, concentric; active coolness. Retreating properties.
Notions of color:
direct influence upon the soul
physical and psychical.
To cause vibrations in the soul:
form, color, timbre.
Multiple depictions of a figure.
Consecutive movement
Lines of force.
Machine as a reductive symbol or a paradigm of energy.
Dynamic interplay: machanomorphic: Duchamp.
Symchromy: Russell.
(Klee): polyphonic painting and music.
Spatial element and time element.
Where the sounds evolve like veritable entities.
Colour sound.
Mathematics of color: Wittgenstein.
De Kooning: Sensitive colours.
New constellation.
Economic warfare
First came Leucocephalus.
Along came Buchaphelus.
Liquid cash
according to M. Ruppert
Cash flow.

Economy. Implosion.
Multinational corporations. Organized crime.
Into the woods.
w.w.leader
accumulated profit
Financial systems. Listed entity. Market capitalization – ratio.
Stock market value.
Economic recovery.
As crooked as the rest of them.
Administrations.
The Dow. The Tea
Infusion of cash
So-called resurrection.
Until the contract matures.
Put option.
Evergreen takes cash over the world´s banking system
Geo-political develop
Extracts from the Grand Chessboard:
Key orbitor. Geo-strategy. Geo-political prize. Global primacy.
To be top-dog in Everasia.
-Subordinates Africa,
-nearer Oceania,
-prevent collutions,
-prevent barbarians to unite.
A deliberate plan:
-Stan-countries. Old Tartaria.
Khan –Land-
Natural gas, Gold, oil.
Pipelines: Who has the technique?
Not only the first and only, but the last = One nation.compr
Chomsky: Direct democracy. Consensus without interference.
System of shaping control.

Has been marginalized . . . thought control.
Avoidance of Time. Space according to the gospel.
Sense of space.
Huxley: From Eyeless in Gaza: "A timeless aquarium, Troubled waters. A true, spirited, inner, higher truth.": "Carefully perverted metaphysics, an ingeniously adulterated mystical theology": (p. 99), "The one is not absent from anything, and yet is separated from all things".
Armored courier. Security froed. Policy to prevent abuse.
High resolution panorama in mosaic: NASA is seeing red.
Genes interfere. Tropical cyclone. Heavenly objects. Black ice.
That´s the timeline,
the happy sound.
Glorified self-disclosure.

January 10th 04

To resist the sword.
The sea of love
Orgon: cosmic rayon; when it hits you, it hits you grand.
Aim, end, goal, means.
Teleological principles
The secret Doctrine, (H.P.Blavatsky).
Stanza iv:
Commentator.
Hierarchy.
Cosmo-psychic powers.
Fire presides over time.
Fire is æther in its purest form. Unity of the æther.
Two fires: : the fire, formless and invisible, concealed in the central spiritual sun.
: Fire as knowledge; those emancipated fires.

The symbolism transliterated, derived from archaic prototypes
Word, voice, spirit.
?Color, form, image.
?Element, number, substance.
Nature powers: conscious intelligent powers of nature. The mystery of sound and speech. The thought, the representative sign is ′self-engraved upon the astral fluid′.
The word or speech is a blessing or a curse, either beneficent or maleficent, according to the hidden influence attached by supreme wisdom to their elements, i.e. to the letters which compose them and the numbers correlative to these letters. (P. Christian, SD, st5 vol 1,p. 94)
The mundane egg. Our phenomenal plane. (96) Numerals are forms. Dots (diameter), lines, triangles, cubes, circles, spheres. Number in its unity of substance.
Knowledge gained when noumenal turns phenomenal.
Sound and speech: effects of Divine thought. Speech movable, mind immovable.
An empty dwelling. Purposely obscure.
Half an eye.
Scintillae, sparks, spawn
The eye is the lock. Phenomenal keys.
As filings follow the magnet
The vital wind (pneuma)
Synthesis of the seven senses: smell, sight, taste, touch, hearing, mind, understanding.
The senses exerting themselves.
The gates of intuition.
The high intuitor.
The magnetizer
Ideation of the mind.
The central wheel.

Energy is the work done by a force.
Anthropomorphic powers.
Real=magical knowledge=occult wisdom.
Eternals and primals.
Truths and causes become omnipotent, as opposed to illusions and false appearances.
Debasing conditions.
(vi, p. 109) Anupadaka=parentless, i.e. self-born of divine essence.
Chaos/Erebos, Nox/Æther, Hemera.
Stz 5,p. 111:Fohat is one thing in the yet unmanifested universe and another in the phenomenal world, there he is that occult, electric, vital power, the transcendental binding unity of all cosmic energies, which under the will of the creative logos, unites and brings together all forms, giving them the first impulse which becomes in time law. As unmanifested he is an abstract idea, the propelling force, the active power.
The light of the superior and the light of the inferior regions.
Propagation of light.
Desires,
Passions
Desire=pathos, as a principle of creation.
In the chain of discourse:
Hobbes: the passions are voluntary motions. Appetite-aversion. Dejection of mind as a want of power. Impudence: the contempt of good reputation.
Endeavour peace.
Violation of faith.
Contract, pact, covenant.
Sovereign, civil, divine.
Miseries and calamities of war. Chxxi: liberty, freedom. Without external impediments.

Hobbes: Leviathan:
Jus libertate: freedom to
Lex libertatis: freedom from
Whether something will be: opinion

Right	judgment	wrong
Good	deliberation	bad
True	doubt	false

The sword of justice is too hot to hold.
Rules out of the practice of time.
H.Spenser: The doctrine of the unknowable. The thing in itself, or noumena cannot be known, but the idea is correlative and therefore postulated. The absolute is equally vague.
Coleridge 1834+: Intuition as the key to reality. Importance of moral experience. Understanding is inadequate.
P. 154: the ultimate principle. The identity of subject and object; a self-consciousness through an act of will, an original act: freedom being a ground of philosophy.
Phenomena: apparition. Consciousness, awareness.
Eudaemonics, aretaics. A sophisticate's catchwords.
Noumenal chivalry. Noumena turns into myth, mere imagination.
Developed enthusiasm. To gather the fruits of knowledge.
An ascending ladder of how many steps?
(Novalis): knowledge is definable as eudaimona.
Intellectual forces.
Virtue which encourages good work.
Mystic city.
(Arh. Waite, Ars magna laonmorum, Wings, 1970): folk elements, collateral fables.
Sacerdotal dance.
Sarcophagi of betrayers.
Flickering lights of symbolism:
scala coeli: the just man is the ladder, the feet on earth, his head

in heaven.
Emblematical.
Through the mental alembic,
veiled in symbols.
An opened entrance to a closed palace.
Pillars, cornerstone.
′A rose is a rose is a rose′. Gertrude Stein
The rose: a bloom of five pedals.
Eversacred and mystical flower of love and talisman.
The symbols melt like shadows in a dream.
The rose relates to sacraments of joy, as love. Radiant, enchanting, innocent, prior to the experience of good and evil.
Not to profane the mysteries.
No rose is without a thorn.
A crown of thorns or garland of roses?
Garment, shell, cast off clothes.
Harpocrates, the god of silence. The rose of silence, as virgo intacta, who invariably and only breaks the zeals. Talisman or preservative against intoxication.
The proper office of all symbols is to increase significance of those myths it is connected. Old myths are truths under a veil.
Those who were never in the land or sea of legend or romance are familiar with the beginnings, but have no eyes for the end.
A fool and his folly.
The Sancgreal.
Toil, adventure, quest.
Perfunctory research.
The holy graal, the bowl of plenty or the secret vessel issues directions, gives oracles, provides . . .
A lance with bleeds at the point
a fractured sword, or destined to be so.
A dish of plenty.

The technical question to attain the end. Valiant, wise and true: fatality follows failure.
The last drop makes the cup run over.
Sacramental and inward refreshment. Cups and chalices in tales of old, reliquary of the precious blood of Christ. The graal as a relic of the passion.
Other hollows: crown of thorns, nails, hem.
Of high symbolic importance
An eternal pantry and cellar.
The keeper of the Graal and his chivalry wear the semblance of life.
The novel.
The Dialogic imagination. (M. Bakhtin: univ of txs prss, 1990, Ed: Holquist)
Inadequacy of literary theory when it comes to the novel.
Polyphony. Marginal genre.
Social, discursive, narrative asymmetries.
Novels lost in character analysis, ethical high-mindedness and watered down psychology.
The morphology of narrative, that is the language of the plot.
(Holquist): the novel is whatever force within a literary system to reveal the limits, the artificial constraints of that system. Always it will insist on the dialogue between what a given system admits as a literature and otherwise excluded texts from such a definition.
Satires, drama, poems become masks for the novel.
Lukács: novel is the characteristic of an age of ´absolute sinfulness´.
Homeric and biblical texts are not represented in contrast since both are, as it were, absolute language, and are foreign to the inadequacies of the novel´s language.
The epic, for example, has completed its development and been

antiquated. The novel, on the other hand is organically receptive to plastic possibilities. The novel is, as Bakhtin repeatedly says, the only developing genre, and was not until recently taken into account in poetics, though in an albeit way.
The novel has accumulated with other genres of literature, incl. art forms marginalized, letters, diaries and so forth.
(5)The novel incorporates other genres into its structure, reformulating and
reaccentuating them.

De Saussure, 1966+

Typological distinction between synchronic (unitarian): language as a system at a given time, and diachronic (developmentalist): linguistic entities studied through time.
Chronotope, artistic, historical.
(85) Simultaneous existence from widely separate time.
Fictional time and space. Adventure time.
Layers
Phases
Stages
Imperatives of the mind.
The source: the healing water.
Fons aeternitatis/juvenalis.
Awareness of the soul: two elements of thought?
The center of thought: body language.
The mind is not an element of thought, but its center. The brain, respectively its organ.
The mind then a part of the æthereal body: immaterial.
Double-world theory.
The soul parts at a given hour. The consequences: The faculties of the mind, as reason, intellect . . . . Whether the soul has

aesthetic qualities; different values attached:
goodness, kindness.
Faculties of sense: feeling, passions.
A sensual soul.
Other worldliness of the mind. Sense and perception are used as synonyms up to a certain point. As sense-perception.
Bodily sensations: (Armstrong, D.M.:Rldg&kgnpl,Lndn 67)
'Feelings: transitive; heat, pressure, movement:
intransitive: vibration, pain, erotic feelings.
Visual sensible qualities: events that are immediately perceived by sight are colour, light and shade, shape, size, motion and spatial relations.
We contrast what we see with the way things look and what we hear with the way things sound. Perception by touch v perception of a bodily state.
Memory and mind
Only implications
No complications
Open society
freedom
rejoiceth but was not supposed to last.
King of birds. The bird being a synonym and symbol of angels.
Entering the Bird's nest: fluttering wings.
To embrace eternity:
the inner eye swells and expands.
In front
buried deep in the head.
At the back of the head: pineal gland. Attached to the back of the third ventricle of the brain: sella animis.
The atrophied eye. The vide infra has been transformed into a simple gland. Yet the index of astral capacity.
Vol 2, pp. 290-300

Thalamus/thalamanchephalon
Cerebral hemisphere
Cephalic vesicle. (*corpora quadrigemina*)
The odd eye has become passive.
The body is the irresponsible organ. The spiritual being sinned.
The karma, (that which establishes the law of ethics); as a synonym of sin is wordly, selfish; desires, passions.
Endowed with reasoning powers:
blind faith or philosophical belief
Windings of perishment.
Labyrinth of life.
Feelings anthropomorphised. An image of the word.
Trust in providence.
Providence is, according to Blavatsky, in the plane of illusion. A whimsical cruel tyrant.
Euhemerized god. Extinct.
A simple 'yes' meant everything. All the agony and pain, and came to symbolize life itself as it lay unspoken in every breath taken.
Angelic sin. (T.Aquinas. Tome 1, Questio 63, artic 1, Summa bonum)
Utrum malum culpa possit esse in angelis
. . . peccatum non praeexigit ignorantia, sed absentiam solum considerationis eorum quae considari debent. Et hoc modo angelus peccavit, convertendo se per liberum arbitrium (free will), ad proprium bonum, abseque ordine ad regulam divinae voluntatis.
Justice as fairness. Rawls:
Particular conception of a person and first principles of justice. Basis in political reasoning and cultural. Conception of freedom and equality specified by principles. Political liberties and values of public life. Ideal social cooperation. Intrinsic association and practices. Extrinsic: law of nations and order of nature. A

moral power, the conception of good and capacities of the self. Maxims, autonomy.
A method of ethics specified by first principles lead to judgment, aiming at good. Sidgwick: The categorical imperative is merely a formal principle only useful in similar circumstances, i.e., relevant and inadequate. Principle of equity. A principle of rational prudence. The principle of rational benevolence. Highly utilitarian. The good of one is of no more importance than the good of another: maximal happiness.
The aesthetic state, Chytry:
The German-Greek love affair. (Grecophilia). Winkelmann, Schiller and Goethe turned the myth of aesthetic Hellas into a concrete force. The small state tied up to its idealist phase. Hölderlin, Hegel and Schelling entwined aesthetic thought in their writings, focused on beauty and myth accordingly. Chytry designates Marx, the progressive Wagner and Nietszche's cultural crisis as realist in the approach to traditional aesthetic politics. Heidegger followed the general pattern of German Hellenism; fracture and images of the polis. Marcuse: counterculture and nondomination. W.Spies: the ethos of aesthetic judgment. The aesthetical state had become a political task of a nation. The aforementioned all in all occupied the same landscape; mythic temporality. Their political fibre accumulated in the aesthetic state. The tragedy of Polis. The judgment of Paris, Solon´s task of legislation, Agamemnon´s sacrifice; a cursed house, the justification of Antigone, the tragedy of Oedipe: such the aesthetic tradition and factor. A political theory of epical, historical and rhetorical bent.
The ´akeda´ of Isaak. (Gen,22:7) Ubi est victima holocausti? Deus providebat.
Different readings: that it was a test. That he was to learn something. The pains of learning. That he was manipulated,

manuduction. Abraham was supposedly not governed by custom, honour, fear and the like, nor torn apart in anguish, but was going to bind his son only for the love of God. He ascended Mount Moriah by his own intellect and will or wanted to make this stupor out of insanity. Another view has been labelled after Kierkegaard (magister ironicus) as a teleological suspension of the ethical: the individual relationship between Abraham and his God. Faith running counter to reason. Highly paradoxical, if you like, as this act is condemned by ethics, but conversely necessary for the religious man. His intention was to do His will and deed. The sacrifice was to save an inhospitable idolatrous and incestuous city; he was therefore venerable. Another attitude is roughly about the impossibility of such a demand. Abr: I will walk in my integrity (Ps: 26:2). Nevertheless. Job: the unhappy love. God might have done so to prove a point to the daredevil, thus tormenting Abraham with his demands.
Samuel Clarke: Abraham could have sacrificed his son because reason told him, whatever the improbabilities, God being omnipotent. R. Jenkin, 10: Abraham knew God too well to believe he would. Utilitarian; to sacrifice one for the many.
The paradox: that the individual as the particular is higher than the universal. (Objective identity opposed to subjective unity?). Somethingological.
-Fact or fancy
The royal, qualitative, quantum leap.
When Parmeniscus lost his ability to laugh in the cave of Trophonius.
Melancholy: hypochondria.
Le mélancholique a le sentiment du sublime.
Sublime: old age. Beauty: youth.
A dying soul´s nausea. Nausea as a subjective shipwreck.
Nostalgia.

Curators. The abyss.
Streams increased or decreased at will.
All depending on the outcome preferred.
Density, fervent heat, pressure.
Furiated sanity.
The lion and the dragon: strength and ignorance.
Lose energy, gain time: heat and validity.
Made, delivered to eye misdemeanours.
The cowl does not make for a monk.
A fitting and proper irony.
Control through sex drive.
Cut through senses. Immediate overdrive. Rigid control.
-They got you from the start. Breakdown unavoidable.
Love comes tumbling down.
Electric memory lane: heart.
Bitter
rough   no go
harsh
Angler´s bait. A touch.
The snail and the strings. Muse upward.
Double clock theory:
no vindication
Money, key, threshold.
Money is energy.
Wisdom: knowledge and judgment.
A group study:
acupoints to control the body.
Realisation>image>reflection.
Density, spatial.
A reflection of a vision.
To recollect sight
realized and imagined.

Re-established forms bound by colour and light.
Electricity, magnetism; made real.
New words reflected.
Real thought (bound); mind and movement.
Imaged thought (unbound);
    material, empirical.
Real thought and stability. Mental force faculties.
Creative recreation of previous awareness.
Altered memory. Recreational area.
Things and thought: space.
Movement and thought: time.
Material or real time.
Real space: color, light and such..
Sight and sense: sense perception; colour sense.
The sight and the eye
Tribochane. (esoteric licence).
Spiral; real.
Cubic; final. Categorical expansion.
Existenz and thought.
Real thought and existence.
Philosophy of right and the lower hierarchy
as opposed to the open society.
Market place arena vs. theatre state.
A crown of thorns is a crown.
To ensure a result.
Suicidal causes.
Animonies of the lost and lonely and the misled.
By ignorance,
addiction,
command.
The sacrifice.
Demand. Command. Tree without a virtue.

The choice.
The deed and the need.
Victims and a sacrefact.
Who´s best suited for the job?
Existential craving: the sacrefact thus chosen.
Forærgelsens mulighed: comic irony.
Ironic fate.
Ironic act. Wisdom superior.
Ironic postulates.
To mock ignoranti, show possibilities.
Subtle irony.
Rhetorical irony.
Real: irresponsibility: afterlife jokes; the stand is ironism.
Cardinal virtues
Obedience, chastity and poverty as medieval postulates
The starry seven: chastity, purity, courageousness, integrity
Submittance, temperance, humility.
Within the province of knowledge.
The will to power vented on thoughts.
Irony and mimetos.
Mimetos and memory.
Platon
The Rep, ix 580DE: Three faculties of the soul
It is worth recalling that the soul is tripartite.

Guardian	financier	philosopher
Pleasure	appetite	control
Victory	gain	wisdom
Experience	discussion	intelligence
Creation	mimetos	<God>

Platonic virtues:
reason
courageousness

temperance
justice.
Varronis eighth and ninth:
Architecture and medicine.
A pure coincidence.
Kant, +1804
E n s
originarium
Transcendentalem
ens summum
Theologia rationalis naturalem: summa intelligentia
Moralem: summum bonum
Theologia empirica
Califer loaded with sultáns:
authority and power
Aporias:(dead ends).
Status originalis
St. corporalis
St. gratiae
Faith, hope, charity.

Mars 15th 04

Theory and model philosophy. Weighty studies by specialists. Labyrinth of professional argumentation filled with digressions, polemical asides and disciplinary boundaries. Nuanced analysis, logical strifles and critical points to weave together scattered threads.
The principle of unceasing movement among principles.
Values and meanings,
change of meaning.
What this or that statement means is merely a technical

question,
quantitative, valuative . . .
Devastating
the fate of our time; the disenchantment of the world.
Separate: agony, pain and the like. Physical passions.
Compare: artificial intellicence.
Inaugurate: feelings-sensations. Faculties of the soul.
No drive.
volition deteriorated.
Ultimately: higher wisdom.
Perpetual loss
Sense of joy etc: no personal sensation any more.
Sight: no inner visual data, let alone sense data:emptyness
Memory: past sight and sound unfound, unless by aid.
Developed personality: existential or essential
Thought: No personal, individual thought. No memory base. Past sound and sight being controlled. No use trying to retrace yr memory lane. Recreated by us. Past feelings thus overrun. (disturbances:statues and soundbarriers).
Mistakes.
Steadfast belief. Unnessesary act. Out of prejudice. (Ignorance). By miscalculation, misinterpretation. Although it were an act of love.
Remember, you´re loaded: unbroken time sequence; lonelyness.
Memory base locked. Change of thought.
Behaviour
Remembrance
Sensational impact depends up on us. We have to agree.
Your conscience.
Individual conscience.
A conscientious mind.
To have a clear conscience. Is it the same as being venerable?

A fable? Like having paid the due? Being, becoming, constant flux. Where nothing bad can harvest.
Every act illuminated. To shed a light.
Sub consciousness. Non-verbal?
′Always something at the back of my head′.
To fear or like something apparently neither from memory nor thought. Something repressed or forgotten.
An inner voice, external command, a whispering conscience.
Self: individual
Identity.
Persona.
Character.
On repressive moments.
To forgive and be forgiven.
Possessions and impact on body:
vibration, attraction.
Energy flow, disturbance, disruption
Psychological effects, physical, organic.
A constant stimuli
If the individual chooses to be, so to speak.
Kant.

Quantity	universal (unity)	particular (plurality)	singular (totality)
quality	affirmative	negative	infinite
relation	categorical	hypothetical	disjunctivesynchr
modality	problematic	assertoric	apodeictic

Nature.
Empirical self.
Transcendental self.
Categorical reading.
Ideas:
Impressed

Expressed
Abstracted
=
insited
innate
reciprocal (area)
Ontological meaning
Principle of Leibniz, thus mathematical meaning.dash
Lying, cheating, fooling
Amicitious, beauteous, orniteus
Five seasons a day
No Dulsinea, no windmills.

April 9th 04

-Trust your mind.
You can look at it this way: your loss is our gain.
The secret of painting: to paint with your soul;
turns automatic
A fairy for children like me. (The child is the father of man).
-and then you will marry.
The tongue
-saliva factor
C-fibres in the brain.
The instrumentalist.
Played by an individual with a consent
=authority from god, if you will.
A golden key opens every door

Mai 4th 04

LW: Zettel, 233: *Das intendierte bild.*
Intend to intentionally. Intended by.

To be made to see what has been impressed or insited.
Sense data and plissement.
Intended and shown, thought and mediated.
Primarily expressed or primarily impressed.
A third anchor.
Taste tripartite: impressive, expressive, abstracted.
Perception, (sense) or feelings.
Sensual words, sentient minds.
Virtus creativa.
To twist the words of Bishop Berkeley:
Percipii creare est.
Very archaic!
Creation. Invention. Recollection.
Create feelings, works of art. Invent laws of nature.
-Every free, perfect or true perception (incl fac of the senses), is a creation. Understanding and reason then invented or recollected.
What happens in the moment when the mind recollects the creation of the senses?
To be created upon. Under constant creation.
A creative act may be collective; perception or sensation being meditative.
Taste and reason

Mai 5th 04

Rainbows and parhelia.
Rays of light or undulation of air.
Process. Transiton or concrescence,
empirical or transcendental.
Leibniz.
Light being kindled:
Essentially existentially

Necessary contingent
Eternal truths x
Universal singular
Intrinsic extrinsic

The whole universe in its perfect concept shall come into existence. Every physical action changes denomination. A butterfly effect.

The hypothesis of concomitance.

Modes, (not things) of space, time, extension, motion.

Ensouled beings are transformed. The government of things.

Leibniz. (Discourse, 254-7): of all beings, those which are most perfect and occupy the least possible space are spirits, whose perfection is the virtues.

When we consider the connection of things we see the possibility of saying that there always is in the soul marks of all that had happened to the individual and evidences of all that would happen to him and traces even of everything which occurs in the universe.

Transformations. Representations.

A substance bears a resemblance to an infinite perception or power of knowing.

The common run of thought.

The clock and the clockness of it is enough for the man who buys it.

A physicist or a geometer goes out of path if he incorporates substantial forms, employs the corporation of God or perhaps of some soul or an animating force in his hypotheses.

Individual substance > substance form

The predicate is present in the subiect.

Leibniz. (Monadology) A simple substance. Monads, Entities, Entelechies, Perfectihabies.

The immortality of the soul

#17 Supposing that there were a machine so constructed as to

think, feel, perceive, and we could enter it as we do a mill. There we would find only pieces that push against one another but never anything to explain a perception.
#19 All simple substances or created monads have a simple perception but as feeling is something more, those substances shall be called ′souls′, whose perception is more distinct and is accompanied by memory.
#29 The knowledge of necessary and eternal truths furnish us with reason. This is what we call the rational soul or spirit.
Acts of reflection: the self, the ′I′.
#47 God alone is the primitive unity or the original simple substance, in him is power: the source of all, knowledge: detail of ideas, will: which effects changes or products according to fitness or the principle of the best.
Every monad is in its way a mirror of the universe. The soul also represents the whole universe.
#67 Each position of matter may be conceived of as a garden full of plants. But each branch of the plant is also such a garden. #73 What we call birth is development or growth, as what we call death is envelopment and diminution.
To pass to a larger theatre: Sensitive souls are, by election elevated to the rank of reason and to the prerogative of spirits. Sensitive or ordinary souls are living mirrors or images of the universe of creatures but minds or spirits are in addition images of the Divinity itself.
. . . To those who love and imitate, as is meet, the author of all-good.
Blavatsky, *Isis Unveiled*, I, 277-81
A docile tool in the hands of the invisibles –beings of sublimated matter, hovering in our atmosphere to inspire those who are deserted by their ′genius′.
(Agrippa, C, *De occulto philosophia*): the soul of the world,

the ever-changing universal force, can fecundate anything by infusing it in its own celestial properties.
Human soul possesses a marvellous power . . . she can shoot through space and envelop with her presence a man no matter at what distance. (Eliphas Levy): : By combinating with the subtlest fluids, the envelope of the soul forms an ethereal body or the sidereal phantom, which disengages only at the moment of death.
The cradle.
Platon. (Cratylos).
Rivers, streams.
carries and holds nature. , refined, is the source of life. When the soul is denuded of the body, (μ) it is bound with the disire of virtue. Socrates: Heraclitus is supposed to say that all things are in motion and nothing at rest: he compares them to the stream of a river and says that you cannot go into the same water twice. (402a)
Hades was called by his legislator from his knowledge. Hera is the lovely one ( ). Possibly also only a disguise of the air, ( ).
The name of Apollo, the god of harmony, is most expressive; embraces music, prophecy, medicine and archery.
Artemis is named from her healthy nature, (μ), or perhaps from ( ).
Δ is giver of wine,
Aphrodite, born from the foam, ( ). (Hesiod authority).
Athena. As Pallas derived from armed dances, elevation above the earth: ( ).
Interpreting Homer, Athena is and , that is mind and intelligence. The maker of names might have been thinking about a divine intelligence: ( ), or she who has the mind of God, (Δ), or even ( ), altered

by successors into Athena.
Zeus Creator may be divided into (and Δ), the two signify the nature of the god and the business of a name is to express the nature (396).
Names with natural fitness: Sosias: saviour, *Theophilos*: the beloved of god.

Mai 6th 04

Sinister,
Cynical.
Bakhtin (The dialogic imagination).
With the sharp of his sword, pp. 193-206
Rabelais, Gargantua and Pantagruel: The Hog of Minerva.
Death robbed life on earth its value, considering it perishable and transitory; death deprived life of any independent value, turning it into a mere service mechanism working toward the future eternal fate of the soul beyond.
The entire situation is a wicked parody on the conception of life and death as it was perceived by the medieval transcendental world view.
In the ′Isle of the winds′ the souls pass out through the ′rear passage′.
A ′dry mass′ with a Good Anjou wine. The holy chapel as the convent kitchen.
The Immortalizing.
From the episode of the Abbey of Thélëme, Book I, Ch 52+
′Thanks to the propagation of seed . . . my soul shall quit this mortal habitation . . . since in you and by you (God) I shall remain in visible form here in this world of the living′.
For Gargantua, according to Bakhtin, it is not important to immortalize one′s own ′I′, one′s own self. What matters is the

immortalizing. ′And thus if the qualities of my soul did not abide in you as does my bodily form . . . I would see that my lesser part had persisted, that is my flesh, while the better part, as it were bastardized′.
Death, as Rabelais expressed in his book, says Bakhtin, begins nothing decisive, and ends nothing decisive in the Collective and historical world of human life.
A conscience (Hegel) which is a particular individual distinct from general substance. Seen from the side of the substance, the consciousness is as a self contained reality but nonconscientious of it self, being now divided against itself.
Locke: vitally united to that which is conscious in us.
Hume +1776. (Concerning human understanding., sec XII).
"By what argument can it be proved, that the perceptions of the mind must be caused by external objects . . . and could not arise from the energy of the mind itself, or from the suggestion of some invisible and unknown spirit of from some other cause still more unknown to us?"
The idea is reflected upon the mind, whether compound or simple, from an impressing.
Opening closed galleries.
Hidden science.
Robotize.
Healty workforce
prosperity
forward go.
Propagation of people.
The process room
-the pain you are receiving
A hierarchy comes tumbling down.
Anarchy and chaos.
It all comes down to this: reign or be reigned.

We´re on to it for you.
Break a clue.
In case of panic
Calm your friend
Go to the nearest station, report.
War terms.
Enemy abuse.

June 15th 04

In the actual.
To bear the causal past in mind.
The mental entity is the process
underlying every human action
Erring entity. Ventured personal. (No records).
Fallacies of thought. Lack of insight.
Imaginative leaps. Accidental influence. Connection with other individual entities. Asynchronic. Hypothetics of the past
The collective.
A troubled collective. As a special state of mind. The wrong stand. A supposed correction: bad reasoning.
The passive become active.
A principle of fallacies: : Every Correction is, so to speak, with a generic leap.
To duce from an unknown entity. Passionate.
Entity and persuasion.
True propositions applied to personal circumstances. Fallaciously led. Might repress common sense.
The empirical and experience: senses. Impressions, datum, occasions, memory.
Mental perceptions cause feelings.
The instance of communications,

the manner of performance
To be put in a state of mind
A feeling of conviction which brings action about.
A.N. Whitehead. (Process and reality, p. 123).
An enduring entity with a personal order. Actual entites, (as atomic), non-dividable.
No physical point from which to draw.
The idea thus developed: the ´who´s to know´ factor.
Hybrid (impure) prehension.
How the subject (the prehending entity) prehends a datum
A premature idea,
impression out of hostility.
Persuasion
Possibly acted upon.
The aura of revelation.
Going fashionably away.
Opportunity seized by forgetfulness or indolence.
From the saucer:
passively received.
Made to act,
(no consolation).
the actual used to
hide a mental dilemma:
spheric unamnity
Substance
Profoundity
Sublation
Sublimation
upheaval
From substance-dualism to dynamic process.
Nexüs and guilt.
Family

Cluster
Group
Society
Class
Substantial entity on a highly categorical/factoral level.
Compare.

Entity	reason
Nexus	consciousness
Prehension	(proposition, statement?)

Nietzsche. The arduous torch.
Living like a true pupil of Ho Skoteinos: in the constant flux of ever becoming.
Looked down upon from the garden of knowledge.
Brought to consciousness.
To describe a feeling.
Man cannot learn to forget, but is forever chained and attached to the past.
Strong instinctual feelings:
moment.
Enthralled by the moment.
The calm of later moment.
The threshold of the moment.
Living unhistorically in blissful blindness, free of invisible and sinister burden of the past. Happiness: being able to forget.
A single delicate injustice.
The horizon is closed.
Horizons drawn by others.
Amidst encircling clouds of mist
Blind to dangers, deaf to warnings
in a dead sea of night.
Goethe: the man of action is always without a conscience, only the observer has one.

The superhistorical man, nauseated, sees no salvation
in the process.
Whether happiness or resignation, past and present is one
and the same.
Three kinds of history:
monumental strife,
antiquarian preservation,
critical liberation
The temple of history.
In the span of time.
The drama begun after the fifth act.
A final closing.
To live at all cost.
Ruled and guided by a higher power:
the sweetest fruit of this bitter growth.
The propelling force.
The anagram of life and death has brought you closer to real existence. Thus propelled, the man of woe, the un-doer of man will strife eternally for a greater cause.
The enigma of great writers comes from above. The lives of the great diriged to the minutest detail.
-(you can do wonders, man. Just think about to join us)
The wonders you´re about to witness, your flash of lightning, as it were, revelates the milestone, the path and the ascending ladder.
The cure you are receiving lies in the hidden context of not wanting to fail.
It is propagated down up on you, matter of factly a life-giving force. Your burden will be taken away in the grace to forget and forgive.
The light of salvation.
It is not where you begin, it´s where you end.
Word and image.

Poet mystic
Word silence
Morale God
Knowing in order to love.
Affectionate
Speculative knowledge
Substantive
Speculative knowledge=experience?
Showing affections in which manners serves you good.
Miscalculations.
Questions about methods.
The receptive approach.
Influential people affected upon..
An active magnetic force, if you like.
Mental weakness channeled.
To clear/rectify other consciences.
Canal for a certain behaviour.
Qualities at odds. Negativity. Vulnerability.
Every sickness is a lack of something.
Sickness comes short.
The tragic guilt becomes real in one person, family . . . thus turning him/it into a victim and a sacrefact.
Where to draw the line?
By the deeds perhaps.
Individuality.
Character.
Adverse guilt.
Comic innocence.
shaping of a personality.
Upbringing.
Resisting powers.
Weaknesses.

Alien influence.
Charmed.
Sensual=voluptous.
Be mindful in the presense of a soul.
Symptoms of the minds bear resemblance
to the physical defects.
Infecting minds.
Housing the deeds of others.
Coition.
Carnal lust.
Dyer´s weeds.
Coloration.
Achromatic:colorless.:
In the hereafter:
conduct of life.
Spark of life
Coexistence.
Phaeton wanted to possess the chariot,
stealing from the fire
Dreams
Poetry
Prophecy
Interior enthusiasm.
intuition and inspiration.
Aragon: the great defeat which perpetuates itself.
At the height of suffering.
Mystical certitude.
The inspired poet and its mystical message
Supercherie.
Charged with the moment.
Words and images.
External actions of the like alongside

with internal sufferings.
Means to an end.
A method which brings forth suffering
to save others.
Magnet for cruelty.
Limit absolute.
Permanent suspension.
Percipient guilt.
Imprisonment.
Improvisation.
Addiction turns to habit.
Lament despotism.
Escalation.
          Image> form
Sign> symbol> picture.
Whitman:
warble your song
envelop´d in dreams
enveloping with the rest.
Philosophy of flux
To think, recall from memory and the like is a mere stagnation,
a SQ, nothingness.
Drives. Passions. Longings. Will and flux.
Existentially individual, social, historical.
Identity and flux.
The will to flow. The vital force.
To see and to express.
To think and thought.
Future induced.
All is flux. Nothing is of importance, neither thoughts
nor actions.
Loving in the flux: happiness.

Before the beginning and after the end.
Intervention causes alteration. Minimalistic. A flux is a flux.
The will of the eye.
Fallacies: misunderstanding or not believing one´s insight.
Repressed feelings tend to change by fright, guilt and such.

July 23 04

Elan vitale,
Compounds.
Force. Conjoin, distract (upheave): intestines, brain. Mass.
Illume. Magnitude, ability to ´dwell´ in many places at once.
Strings to heart
from kidneys, glands, stomach.
Places from which to see.
Movement. Eyes.
Intuition and individuality.
Nietzsche: beyond good and evil. 1886
L´effet, c´est moi.
The body is a mere edifice of collective souls.
Les vitrines de l´ame.
Ernst Mach. 1836-1916
A color is a physical object upon
luminous source
upon other colors
temperatures
spaces
However, its dependence upon the retina: psychological object.
(*The English philosophers from Bacon to Mill*, The modern library, NY, ´67)

A. Compte	Mach
P. 742	p. 763

Mathematics	colors
Astronomy	sound
Physics	temperature
Chemistry	pressures
Biology	spaces
Sociology	times

Mind, feeling, volition.
Ancient, mediaeval, modern.
Mach: permanency of the ego.
Relative: quiet, cheerful, excited, ill-humored.
Apparent: continuity. Slowness of its changes.
709 the little habits that are unconsciously and involontarily kept up for long periods of time.
Frequent bowel movements.
Birth of analysis. Reductionists.
Death as liberation from effective individuality.
The element of change in bodies and the ego is what moves the will.
From the solid to the vapor: sublimation.
Reverse: deposition
Compounds of hydrogen: binary hybrides grouped into three broad categories:
covalent
ionic
metallic.
Covalent hybrides are formed betw hydrogen and nonmetals.
Radioactivity.
Alpha particles, (2 units of positive charge); (same mass as He, identical to He2+ions).
Beta particles negatively charged: same properties as electrons.
3d form of radiation are gamma rays. Not particles but electromagnetic radiation of extremely high penetrating power.

The chemical properties of a radioactive element change. Fundamental changes at the subatomic level. In radioactive decay one element is changed into another: transmutation.

Aug 10th 04

misfortune never comes singly.
Dewdrops.
Libellous poetry.
Revile content.
The fiddle: a titanic instrument.
Salvation through suffering.
End of Sublime.
Defect or weakness.
The causal law questioned. A materialist's way out.
Materialism.
Particles, nerves, brain . . .
no thought without a material activity.
Functionalism: the law of cause and effect is contingent.
Eliminative materialism and flux..
Logie superior
-love is submittance.
Embriology.
The identity of an individual is established epimaternally and epigenetically at 20-30 weeks of pregnancy.
There is one thing to identify, another to have identity.
A genetic identity of an individual? A fable.
Two or three generations ago identity may have been established at birth. There was a qualitative leap.
Identity has nothing to do with population of cells, genetic development or any parental factors.
Living light.

Animal rights
Preserve wildlife/other cultures.
Those who are nesting bad people.
Bad girl pet.
Bad boy toy.
The show must go on.

Oct 14th 04

Moulding time.
What shall the clay reply.
Moulding the soul:
Affliction,
Fornication.
Delivered up to the sword.
The dead and the clay.
The process room. A mediaeval castle. Torture chamber. Past selves brought home through physical pain.
The same witch executed 47 times.
Sin and punishment.
Sinned according to the punishment of the times.
Time of sin committed and accepted. Punishment
of the same hour.
It is obviously not like that. Punishment is decided when light has been shed.
On crime.

Pay in like manner.
Fear of temptation.

Bad luck and injustice.
:hardening of selves
:sin committed in the light of the times
:foregoing punishment

:the council of the wise. Fate thus determined acc to the customs of the time. Their future designed when sin became known
:losses. Things undone
:illusions
:mistakes
:perpetual loss
:tomies
:mutando mutandis
:things change overnight. Emphasis accordingly. Historical turning points
:zeal, vehemence
:lust infused
:shall the clay reply?
:the body the irresponsible organ
:dox of cause and effect
:social life/standards before
:chain reaction; revolving snowball
:probability
:new methods, techniques
:debits, credits
:original sin
:igno-traps

Number, measure, duration.

Auto sacrifice.
Material. Things, money, work.
Friends. In short: a life.
Independence.
Freedom.
Will.
Rejections of the past.
By opening the gland, to draw force natal/vital to the brain.

Personal and individual thought.
Identitate to yr own self.
How words and images come into being/awareness.
Pre-tongue.
Symptoms of an era.
Prisoner´s dilemma.
Structural violence:
vioaltion can take many forms; oppress classes, groups: freedoms and rights,
thus selective.
Game theory. Conflict solutions.
Minimax
:abolish want and poverty:
the power vested in one person.

Nov 7th 04

Feelings overrun. Artificial inauguration to correct the deeds of others.
Snowball theory.
A historical leap.
Individuals imprisoned in the physical.
The architecture, the new-hierarchy, or better: the superstructure.
The binding force.
The ´elan vitale´, the vital force is passed on to a physical object which will in due time ´inherit´ that force and become positively stronger and more apt to take part in the process.
That is to say that the being ´inherits´ this force from his predecessor, this mental alembic.
Exodus, (5:16)
(prisoner´s dilemma)
idle work

evil case
<and I will harden his heart>.
Johannes (10:18)
Power over own life.
The broken commandment:
<Therefore doth my father love me, because I lay down my life, that I might take it again. No man taketh it from me, but I lay it down of my self. I have power to take it again. This commandment have I received of my father>.
I am neither free nor do I have volition. I have not power over my life. The word thus down the mainstream.

Nov16th 04

Castaway.
Making plans for nobody.
Strength, grace and decency.
General misconception..
The structural revolution (structural violence
as on the effects of democracy fx), uses
like any revolution classical methods like:
terrorism
sabotage
propaganda
kidnapping.
It now seems to me that your enemy is a terrorist
secundum quid.
He is first and foremost a revolutionary. To label this revolution, I have chosen to call it a structural revolution. Their technique is as classical
as any, I suppose. The structural violence capitalism exposes, (in fact every governing factor). A revolutionary is a reformist

of some kind. Ancient values? It may be an overstatement to say that a man equals tank. Notice: their religion outwardly says that the afterlife is more real. It so looks to me that as so many of them, (I´m postulating rural, even general support from the masse), wanted this revolution to be brought about, their only warfare would be guerilla-like.
Some questionmarks:
bogus election
woman´s rights
poverty.
Price of lemon-drops is dropping
In line with what you´ll find:
to monitor for the entire period
Reiterate:
toss the dice.
Hit you with lightning
Zeal without knowledge is a crazy horse.
There are things to do if peace is to be established:
stop suicidal acts,
Sstop sabotage,
leave captured territories,
freee prisoners of war.
A drastic shange in relation to Israel and Palestina is needed. The U.S have on numerous occasion used their veto, favoring Israel, and have by their might disregarded the will of the league of nations.
It has now dawned upon us that this revolution won´t stop even though Iraq will be put to ruins. Buchaphelos is made to run. Where will he turn next?
The revolutionaries do not lack generals. If one falls, another one rises. It´s the nature of this kind of warfare. Every guerillero is a general in himself. Then the rural factor. Knowledge of territory,

the oppressors on a hostile ground. It might be the case that one individual is the binding force of this apparently invincible army, but then again, there are too many potential warlords. How many cells? In how many countries? It the majority of the Arabic/Muslim population just pretending to be asleep?
Those who profit from this war are the passive bystanders.
There won't be any secularisation oppressed on any nation. Her heart turns to steal.
The international community is lightning-hit. The burden of the past is building up fast for this arrogant and angry young nation.
Revolution against imperialism. Against any discrimination.
The power of the people. The guerillero is but their instrument.
Indoctrination of the people.
A guerilla warfare, (see fx La guerre de guerilla by Che Guevara), has its scientific notions, a tactic, structure, strategy (terrorism, sabotage . . .), goal.
There is always the effect of surprise.
The rebellious army.
Motives and goals: stay loyal to the revolution.
The moral factor: international sympathy.
Defensive.
:escape routs
:mobility
:adaptability, improvise
:local support
:disarmament impossible
:secured places
:point of determination; moment chosen
:interaction between the public and guerilleros
:camouflage.
A radical movement with a mission.
The guerillero has to integrate with his environment and carries

his house on his back.
Spirit of sacrifice: heroic acts. Optimistic and tough to the extremes.
Rigid scheme.
Method.
The bishop in the play ′Balcon′, by J. Genet: "you cannot do evil. You live in the evil. In the absence of regret. How can you do evil? The devil plays." He′s the grand actor.
Remote viewing: clairvoyance.
Heaven′s den.
Infrastructure.
The front
Create catastrophes so that the price drops and/or create good outlook.
Backup group of investors,
Capital,
money flow.
Hint: sell out/overtake.
The outlook move:
Capital overthrow.
1 increase shares
(to create good outlook)
>price goes up
2 backup group sells
>price drops
3 shareholders lose
microeconomics: to explain and predict the profit, growth, sales of firms, market . . .
private or separate ownership: corporations of limited liability.
Company laws. The act (1933)
The capital.
a) fixed (buildings, machinery)

circulation (semi-finished products)
b) workforce
c) essence of time
d) wealth
Financial capital or productive capital.
Marx: constant: machines.
variable: labour power.
capital property.
neo-classicalism.
Impetus. A force which causes an object to move and not to change its speed or direction.
JWV Goethe, Bd 13, Ch, Bech, Munchen ´81. (Thaschenbuch), #648-748:
Newton bezieht die sieben Farben des Spectrums auf die sieben Töne der tonleiter.
. . . So wir können uns einen Verstand denken, der . . . intuitiv ist, der Anschauung eines Ganzen als eines solchen, zum Besonderen geht, das ist von dem Ganzen zu den Teilen . . .
intellectus, archetypus, ectotypus.
Light machine.
Infra-optics.
A tragic leap.
Auto sacramental one act allegoric play performed in squares on corpus christi day.
Spanish heroes esteeming love and honour who died in a duel, f.x. were condemned by the ecclesiastic authorities, and are thus tragic figures.

Nov 23 04

Economical systems may come tumbling down in the near future. A calculative failure. The reason for this is twofold.

Firstly, the money flow is one sided, (from ´bad´ companies to a sheltering one and out).
Competing the open marketplace has become superfluous to say the least.
A firm to cut corners with, to buy or sell through that firm, thus obscuring the goal, making it hard to trace where the profit will end.
Contradicting theories.
Example:
firm A buys a house located on Lane 1. firm B buys from Firm A. Firm B is owned by A. Firm B has firsthand knowledge about Lane 1.
B +1a (price goes up).
A +1b
1a &1b (price drops).
B+1a
B=1n
Illegal?
B owns a firm. B sells a part in that firm on a settled price, (lower estimated value), to himself ,(another firm he owns). The price drops. Others buy, (others buy; and because of that the price goes sky high). B sells his share=his gain. The firm loses credibility or goes bankrupt.
Stoolpigeon
B owns a firm
He sells it to himself
Others buy
Price goes to the rooftop
He sells out
Price falls again.
B&f
B1+f

N+f
B-f
Attraction/distraction of elements.
By thought.
Electrical charges.
Form: cellular.
b
Light
Sound
Vision
R
Light
Vision
Appetite
B
Light
Sound
Vision
(sense)
Function: b is dependent on R and/or B. R is dependent on B.
Different vision. R and B are similar, but b not as clear.
Different sound. B is baser than b.
Energy mass.
Atmospheric aid: expansion in space.

Des 7th 04

Phantoms in the brain
V.S. Ramachandran (fourth estate, lndn, 1998).
Neurology.
Neurons w. cell body and thousands of dendrites which receive information from other neurons. Each neuron also has a primary

axon (a projection) for sending data out of the cell.
Gross anatomy: that greyish liquidy substance

Motor cortex
Frontal lobe sensory cortex
parietal lobe
occipital lobe
Optic tract
Lateral fissure Corpus callosum
Retina hypothalamus
Pituitary gland
hippocampus
Cerebellum
pons
Larynx
(adam´s apple)
Thyroid
Windpipe spinal cord

Lobes fx separated w furrows or fissure.
The limbic system is neither directly sensory nor motor but info built: telephonic motor-system.
Structures surrounding a central fluid-filled ventricle of the forebrain and forming an inner border of the cerebral cortex.
Chemical messengers, hormones, adrenalines, dopamines, peptides,
neurotransmitters.
Androgen from gland.
Neurons transmit and receive signals. The structures of the limbic systems, such as hippocampus, amygdala, septum cingulate (p. 100-101).
The basal ganglia: initiates skilled movements. Structures s.as candate nudens, globus pallidus and putamen.
>Tract>nucleus>gland?

Sensory pathways.
The control room.
The thalamus may be considered as a relay-system, transmitting info from one brain region to another.
Upper diencephalon= thalamus
Lower diencephalon=hypothalamus: overlies the pituitary and controls its hormonal functions. The HT´s activity partly controlled by hormones released by endocrine glands, s. as eating, drinking, sex, emotions. Also strongly connected to the limbic system.
Lateral geniculate nucleus receives visual information directly from the retina.
Peripheral neurons control the internal organs and glands
105 the reticular activating system is located in the tegmentum of the midbrain: wakefulness by electrophysiologically activating neocortical neurons.
Reticulum (net) smooth muscles. The tectum (the roof) of the midbrain (inferior colliculi): auditory iinformation. Superior colliculi: visual info directly from the retina, but also interconnected to the neocortex.
The medulla lies in the myelenchephalon: regulates cardiovaskular and respiratory functions as well as maintaining skeletal muscular tone.
The cerebellum receives inputs from visual and auditory systems, (from the cochlea of the ear), and sends outputs probably to systems needed to motor control, enabling complex skilled movements.
Weight: 1440gr. 2% of body weight. 20% of oxygen supply.
Linked by connecting tract and commissures. The cortex is the outer surface of the hemispheres. The neocortex is divided for convenience´s sake into four lobes:
Frontal

Temporal
Parietal
Occipital corices of left and right.
But it might be better to classify acc. to its cellular architecture. Can be identified by the info processing they perform. Brodmann: sensory: eye, ear, touch, pain, temperature, sense of movement, i.e., kinaesthesis; motor: sends signals to the peripheral system to initiate voluntary movements; association regions: thalamic activity?
Marginal literature.
René Thom. (Structural stability and morphogenesis, W.A.Benjamin,inc, ′75).
Morphogenesis:
electro-
structural-
mathematical-
dynamical, technical
logical, natural.
Chaotic
Random
Arbitrary
By coincidence.
Structural stability
Occurent stability
Cross-give
Give-send
Give-take
Taking changing
Cutting capturing
Capturing emitting
Structure
System

Form
Systematic description,
terminological
Aesthetical value of descriptions:
geometric description of movement> lineal, circular.
Functions of taking (interacting), giving, transmitting:
irreversible transmitters
The morphology of the process, (p. 38).
Structural process.
Pseudo clusters
Imaginary groups
x > self
x >
hyper surface
plane
structure
/X/ the absolute value
Rigid harmony, resonances.
Measurement: weight/intensity, mass, density.
Vague attractors,
pairs of intersection: transmit-receive
Give-take
Fibred
Infiltrated
Enveloped
Folded
"threshold stabilization" R. Th, (p. 226).
the swallow´s tail: the blast sphere of a shockwave.
Paradigm: rapid-slow
Degrees: greater-less
Avoid-seek for
The model of the simplest epigenesis is called mosaic :RTh

Transformation
Metabolism
Change
Director layers, maps
Boundary field
<Maintenance> transmitters
Fundamentals> manifestations>substances.
division
spheric
extension
Specialisation
Adhesion
hyperstructure as a pre-tongue.
Carriers/messengers
Differences- similarities
Variety.
R.Th, 158 a <copy of information> as an example of a peppered anthropomorphism.
>a dynamical abstraction. Information is acc. to a mathematician a <geometric parameterization> and the copy a <spatial extension>.
Absolute
Complete
Replacement, substitution
Reduction to pure physiochemistry,
organism or vitalism on the other hand
Significant factors,
bits and pieces.
Properties of matter.
To dismantle series/schemes.
Gradual approachment with defect or with excess.
Evolution and continuity.

M. Vygodsky,(Mathematical handbook, MIR publishers, Moscow, #368-400).
To increase without bounds.
Free (rather independent/ sums without restrictions) numbers, movements.
Resemblance.
Reading or depiction.
The integral calculus
developed out of need to find.
Volume, areas, centres of gravity.
Differential calculus=if distance increases without bounds, then it receded to infinity.
Concave
Convex
To be by virtue of something.
Vibration, waves in repetition coordinated, have their points, top or bottom paralell to where the curve cuts the abscissa.
Vortex
Zenith
The zenith of a whirling light motion.
Bend: loop.
The essentials of linear algebra, (´87, Rea n.jersey, dir: Fogiel, Max).
Row and column.
Bell and stick.
Cubic and coordinative (rectangular or oblique
coordinative system: Cartesian system.
See also polar system of coordinates).
Every system of Linear equations has
one solution
no solution
is infinite.

Consistent or inconsistent
S=(x,y)   S=empty 0
Coefficients or constants.
Trivial or polyvial solutions.
Valid by a performing ability.
Sound but not valid unless imaginary value.
Equivalence unbound by truth or falsehood.
A vector space is represented with a
line
plane
two planes
Subspaces.
Inner product space.
Function= association>image.
4.2 The kernel (null space) of a linear transformation.
Eignvalues.
Eigenvectors.
Eigenspace.
Zoological energy.

Des 14th 04

Christians, the utilitarians of the world.
A religion based on a sacrifice for the many
Mi-eco
Security
Person
Musica
Schoenberg, A
Notes, sets, forms
1 6 7 12
Eb C Bb F

2 5 8 11
G Db D Ab
3 4 9 10
A B E F#
To amaze mediocrity
Theory of harmony
Accidental harmonies supposed to be practicable
Dissonances resolve by step downward. The root
Leaps upward.
The chromatic scale as a basis.
Notation:
Rhythm: length of tones: tempo
Pitch: how high or low the tones are
Rhythmic meter
Classem
Morphem form
Lengthened-slackened
Accidentals: sharp flat
Inductively by experiment, listeners came to consort with the wavelength. Its vibration and frequencies: indications and notations of pitch and rhythm.
Inserting of accidentals to harmonize the measurement of vibrations acc to the best light of the times, custom and convention.
Rhythmic value.
Symbols that instruct the reader to play according to the notation system, the metric signature.
fermata (bird´s-eye) A momentary downbeat (lapse) in the metric beat while holding the note marked with a fermata.
The grace notes (printed as a miniature) are without a measure.
Quite a few symbols for dynamic level ,(loud-soft).
Example: f: forte (loud); p:piano (soft) Words like fortissimo

and pianissimo, crescendo and diminuendo are written in abbreviation where needed.
Articulation, style.

Staccato Dots, under or over notes or pointed wedges. Opposely the slur, a smooth connection of notes, sung or played without interpretation. Dashes over notes to indicate lengthening. Notes tend to vary from one tonic to another, its tonality then highlighted. Such a tone culminates, as a veritable find, the turning point in the work, the binding force, the play´s apotrosis . . .
Scales.
Notes relate to each other in the signature as either flat or sharp or natural, all depending on the tonic, the key signature.
The major scale. The tonal degrees have differing distances, their steps thus in inequity.

C d e f g a h c
v v

because of the shorter interval between e and f and h and c, the c-key gains some of its nearest character, (h).
g a h c d e f g
v v

by sharpening f the interval has moved and put g-key on the level with c major. Other scales in like signature.
Each and every key on the scale is thus measured and balanced. Vibration and frequency tuned, let alone modified by a constant.
Other signatures exist of course
Patterns, combinations, construction.
Sharp-flat-natural
High-low
Upbeat-downbeat
Loud-soft
Strong-weak (beat)
Long-short

Degrees, levels.
Euler. Son et ouïs.
Perception de la perfection se reduit a l´ordre
=causa horologium . . . ubi sit ordo . . .
la hauteur, la durée, intensité
deux sons
3/2 > 1,5
5/4 > 1,25

Des 25th 04

In Obeidah, the Edomites are condemned because of their pride. Luke mentions the twelve tribes of Israel.
Zech, the prophet: 30 pieces. Judas I is one of the twelve. Also he was condemned. Obeidah sympethized with the Edomites. Jesus calls Judas a friend in the Gethsemane, Matt: all that has happened. Did so necessarily to fulfil the prophecy. Now with this one is sure to sympathize with J. I.
Demi-organic things.
Intellectual perfection.
Show-transmit.
Information change.
Systematic interaction, (internal bound).
Constant flow.
Every attack craves a defensive.
As transmitting may involve giving,
even without a resonance.
To know by heart that history is bound.
The constructive vision inductively.
Reveals dynamic possibilities. Whether
the interpreter be informed by explanation,
enlightened by a flash of insight, or inspired,

intuitively calculated by whatever logic available,
depicted, descriptively.
A language come to ruins.
The aesthetical devoid of morale.
The ethics of love emitted reversely.
The repetitive resolution. The differential limit.
An absurd reduction:
the ever-expanding moment.
Rarifications. Accidents in time
repercussive, reciprocal. Dimensional interference.
Expressing at will, thus
impressed by the mind.
Constraint. Onto one-self; transgressing.
A sentential message. Actions of restrictions:
the right to choose, then restricted by aversion.
A judgment like that is metonymous
to the very opposite, namely
some kind of an elusive right.
The floral ornaments, (as seen from an horloge).
The technical décor.
(The veritable -out of it-).
Metaphysio-chemical language,
organic, vitalistic.
Construction by entities. Light. Energy.
Utility. Mutual profit attainable.
Aereal transparence.
To establish sequents, structural informative stability.
Intervention of time.
Mood: recreation of a personality. Existential properties.
Desired drives. The defensives, the limits. The sentential,
The semantics of feeling.
The creative thought as an individual awareness.

Reflective or meditative,
come what may: a conscientive stream, in the now of the mind,
charged with the moment; sharing vocation.
From the hithermost standpoint, the extreme summit.
Envelop´d identity revealing itself from a
premeditated string, bound with the past, as it were.
Measured and synchronized, moderatively weighted
and positively stimulating.
A bend to take a stand.
The irreversibility paradox..
Mi-eco
Capital value measured in work, product
>the gurus
Semantic prosody.
The seriousness of the matter. The intensity.
Obtruse disadvantages.
Semantically distinct words with exact meaning,
which refer to,
Denote or signify an object or an idea.
Supposedly giving your existent,
Reversely taking over the clay
in selfishness.
Character is losing time

Des 27th 04

Heine, H. (Kunst und litteratur, p. 155, bd12).
Eine philosophie der Geschichte war in altertum unmögligh . . . : die menschliche Natur und die Werhaltnisse, beide in ihrem Konflikt, oder in ihrer Allianz geben den Fond der Geschichte, sie finden aber immer ihre Signatur im Geiste, und die Idee, von welcher sie sich repräsentieren lassen wirkt . . .

Das Irrlicht

Amadeus philosophicis: you´re the cause, I´m the effect.
An energetic notion of color.
From linear, (spatial), to colors
Colors and energy.
Hot and cold colors.
Absorb light or emit.
Wet and dry.
Density: heavy-light
strong-weak
Overlapping each other, underlying,
entwined together, woven.
Lines of force: volume.
Cusps of time:
the longest day,
secularize the lot.
The grand preceptor . . .
Whims: character defects.
Capriches: bad intent.
Aereal sight.
Change of movement.
Education: morale
communication
A space created in the head of each and every living being: easier to influence.
Mctaggart, (The nature of existence, Ch VI)
The existent has a substantiality, i.e. a nature. A substance then is an existent
Which possesses qualities without being a quality
Simple a subject relates
Compound to its attributes
Complex

Substance
Timeless
Persistent
Describing according to relations. The relation of similarity/ dissimilarity, of diversity. Causal relations.
Without determination. (Representations).
Tarski
Metaphysics deals a fortiori with the basic concepts of semantics.
(μ).
Moore: the nature-fallacy.
Semantics.
  etymology
  syntax
  prosody

Jan 5th 05

Acids.
System shut down.
Catapheric voyage >to hell.
<Epha> Zec 5:9
. . . going round about.
Jerusalem: a cup of trembling.
The burden of the word.
Judgments, visions.
The idol shepherd.
La fee et ses données.
 : perfected
La coupole spleenétique: Charles Baudelaire.
Martyrisé du Temps.
(Jacob)

. . . thou hast power with god and with men and hast prevailed
The book of Joshua, (the divinely anointed representative).
(3:7). and the Lord said: This day I will begin to magnify thee in the sight of all. See also (4:14).
Sam I
Samuel, dedicated to the Lord, blessed. (alongside with his parents). The word revealed. Established as a prophet.
Samuel´s human judgment reproved.
Sam II
God´s messianic covenant with David

Feb 4th 05

Focal air to attract.
Viscosity.
Deliquescence.
(-I have seen a beautifully sculpted brain).
A living water.
The daily bread.
No recollection.
The mere being, the ontological agent is unable to recollect anything.
Mind of minds.
Superlative.
Waves of future past fanning flames.
The technique was developed in the 60ies.
Experimentals to perfection.
The Entelect.
Calvin.
Predestination.
The doctrine of the elect.
The doctrine of living saints.

Sweet cheat.
No ´I´ is truly great.
The strangest language.
Another coming: revelation.
The virgins of the sea
whose hearts of love innocent . . .
Brontë, Emily:
´No covard soul is mine´
Blake, W, from ´Song´:
How sweet I roamed from field to field . . .
Then stretches out my golden wing
And mocks my loss of liberty
When the center does not hold
On leaving vivid air.
Spiritual insensitiveness.
Angels falling, stripped of feathers.
Bound by nought.
Isa 55:1 <To those who cried out for a living God
those thirsty souls have come to the waters> .
Gal 5:17 <the flesh lusted against the spirit, and the spirit against the flesh>
2Cor 7:1: <let us cleanse ourselves from all filthiness of the flesh and the spirit>
The flesh?
Transgression forgiven.
A blessing, no sorrow added.
In life eternal.
The Holy Ghost as comforter
sanctifies
as minister of flaming fire.
Hearts and minds kept through Christ,
who has come to seek and save.

(Acts 24:16), a conscience, void of offence.
Saved by grace through faith,
sanctified through truth.
Hope set before
He maketh angels spirits of power
love
sound mind
(Ps 149:4), the meek beautified by salvation.
The mercy seat (Ex 25:22)
The wages of sin is death (Rom 6:23)
The soul that sinneth, it shall die (Ezek 18:20)
To be carnally minded is death (Rom 8:6)
Sin . . . bringeth forth death (Jas 1:15)
Kindness.
Charity.
Suffer, bear, endure.
Abound in sure and steadfast hope.
Cleave to goodness, be upright in heart.
Your mind confirmed in your image.
Hermetical=cenobytic.
On symbolical locations.
In Summa Theologica, I, 390b, Thomas Aquinas quotes from Augustin, that :cognitione daemonis sunt privati. "si cognovissent, nequaquam Dominum gloriae cruxifixissent" (I.cor.II)
Thomas Aquinas, vid 110a, STh 1,15
Utrum deus se cognoscit per ideam? Respondeat quin ideas in forma velut in cognoscente sunt. Naturale: ignis generat ignum. Ut preexistens in mente, haec ph. Dici, res assimilare formae.
"Dei essentia est idea in Deo"
Per rationem et ideam
Ideae entelecti Platoni attonans
By animation of concepts past, the need for a new terminology

comes obvious.
Cause comes from upabove, effect out of act.
Allitterative letters.
The speed of light is an extraordinary independent quality and a good clock.
It is independent of the movement of the matter, intensity, color.
An empirical fact.
The criminal mind and its bodies.
Porn hides various social conduct, s.a. drugs, rape, beating, even killing.

Feb 15th 05

Substitutes.
Stripes not directly from the retina, but from a tube there-of.
To take in units. The units then connected to some musculatory system to verify whether an individual is behaving rightly or not.
Your energy, your gift.
Sentients
Entities
Units
Individuals have sentients.
Placement of sentients.
Owner of unit gets no signal/false signal and interprets that as no hindrance: hooked.
Units made to enter in numbers through the pupil.
Image chosen. Individuals hooked.
Something´s sure to go wrong if the poet and the hero habit the same breast:
the hero slayed the dragon but the poet never left home.

Feb 21st 05

Politics.
First the greedy then the needy
Treasure hunters (in the name of the government)
Elections based on false premises
Lives sacrificed on the altar of stupidity. Indifference about the lives of civilians.
Lying to the public:
    what the war was for.
    about the death of O. Bin Laden. Him being alive.
He actually died the last week before Christmas 2003.
Blind w. hatred, disillusioned pride
Treatment of POWs.
Trespassing neighboring countries > hostility.
UN held back.
Villages turned to ruins. People´s homes.
Desecration of the soil, religion, institutions.
Encourage civil war, provoking it even by the very presence
Foreign affairs and objectivity.
On Muslims.
The German way.

Fundamentalists   liberals
X   cultural fusion

____

0   free to join the Democracy
Theology
The courtroom of J.Ch.
Nefi, 1, 18:2, The book of the Mormon.
I built the ship as the Lord hath shewn me, and it was therefore not made for the ways of man.
The material, the ship, the pantry, sailing, bondage, Lord´s compass.

March 3th 2005

The joy, then, is to meditate
An entelect:
a being whose physical abilities are made to be used to give.
An entelect´s life has been perfected. His mental abilities stem solely from the afterlife. His physical, (eyes, muscles, glands, feelings, etc), are controlled.
A being, whose ´life´ is controlled to the minutest detail for a greater good
Why?
By choice.
Create guardians
Because there is need. Evil cannot be tamed.
His physical force is such, that mistakes,
bad deeds, etc, can be corrected by numbers.
Thus the clay has been hantered with.
How?
Afer studying the human body for milleniums,
the hereafter has developed the technique:
the chosen few, the workmen make their weapon.
　　　Light. The flaming light
The human brain and knowledge thereof made it possible.
Now, techniques have improved in the lapse of time.
Gain: manifold. Healthier workforce, higher moral,
crimes, sins and ignorance fought with more vigilance.
To sacrifice one for the many . . .
Dialectics
Escalatoric
Categoric
Perception
　　　Sense

Feeling
Movement from a physical feeling to mental sensation, it´s moment; perception.
Creation
  Recollection
Invention
Movement from recollection to invention; the creative moment
It being. Ontic.
Its ownmost mineness.
Giving
  Being
Having
One might give both meaning and identity a metaphysical weight, but in respect to mind and being, both of these concepts have to be understood semantically.
Identity in itself relatively categorical.
Kant: transcendental, noumenal, as befits on conscience: the soul, and god deliberately knotted out.
God´s interior.
Your voice is pre-emptive,
synchronized
Lying conscience
Cradle gifts
Morning gift
Blood miracles
Physical receptors
Love as ethereal good and the goodness of the mind
How does love come about in the flesh?
The screen
The platform, the ceiling
holding the pillars.

Mars 7th 05

Anatomy.
Cells; growth, movement, reproduction > tissues> organs
Interconnecting channels.
"Rna acts as a messenger for DNA instructions to the cell about activity, repair and synthesis": a codon.
Central nervous system>sensory input from the skin (touch, pressure, warmth, pain).
Nervous tissue.
The human nervous system internal and external stimuli- impulse caused by electro-chemical agents.
The blood tissue.
For transport, discharge.
The spleen located high up in the abdomen: involved in creation, conservation and destruction of various blood elements
The lymphatic tissue.
Function to drain Lymph-fluid formed in the peripheral tissues and to manufacture lymphocytes which are involved in body defensive process. White cells: granulocytes-defensives
Red cells: erythrocytes: affinity for oxygen.
Blood: cytoblast?
The bones also a mineral reserve for calcium
The locomotor system.
208 (?) skeletal bones.
512 (?) separate voluntary muscles.
function of bone and muscle tissues.
The new Atlas of the human body, (Ed. V. Vannini And G. Gogliani, Corsi, ′78, p. 19) "Acetylcoline released by the fine nerve endings crosses the gap to cause depolarization of the membrane. A spreading wave of chemical change causes actin-myosis. Molecular links to shorten, and the muscle contracts".

Touch receptors.
Position “ in muscles and joints.
Stretch “ coordination of movement in the cerebellum.
Muscles.
Not only locomotion but assists in the circulation of blood and protects and confines the visceral organs. Provides also the main shaping component of the form.

Procerus
Depressor supercilli m
Orbicularia oculi m. (palpebral part)
Nasalis m.
Levator labii

The circulation.
The cardiovascular system. The hydraulic (pulsative) pump pulses the fluid through arteries and arterioles (tubes), capillaries: act as exchange vessels, oxygen, ca. diox, water, ions, various metabolic products pass through the capillaries.
A collecting system of lower pressure, (venules and veins), deoxygenated blood: separate circulation via heart, through lungs. Oxygen continent restored, carbon dioxide reduced in expired air.
Venous return and cardial output.
To provide oxygen essential for metabolic processes and discharge. The body of carbon dioxide.
Portal systems, e.g. liver.
The heart.
Systemic,(deoxygenated), and
pulmonary ,(oxygenated).
Endocrine system.
Their secretions,(hormones), pass into circulation.
Pineal
Pituitary

Thalamus
Thyroid /parathyroids
Thymus
Adrenals: just over kidneys) Paired glands. Consist of a cortex and a medulla with hormones. Affect metabolism. Gucocortioco ids>Carbohydrate metabolism.
Mineralocorticoids. Such as alosterone>sodium
And the sex hormones: androgens, oestrogens.
Also: as a result of stimuli releases: adrenalin and noradrenaline
Pancreas (abdominal gland): exocrine, endocrine secretion enzymes > produce glucagon, insulin. Carbohydrate metabolism?
Testis(ovary)

Symbols $\frac{a}{b}$: to simplify computation.

Economy of orders.
Energy of motion.
Power industry.
Electric power.
Nano
Molecular electronics
Opto
Mechanical
Potential energy
Kinetic
Math is power

Mars 23 05

Children

Anger
Fright joy >interest
Pictures
Music
Words
SELF
Family
others
want
not like like
don't
want
concepts scaled.
Motion gets soon enough sentential
which comes to be a part of your soul:
receptive, transmittive: co-acting individuals.
Sentients and emotional reply.
Some sentients tend to grow. They form a cluster, (an organism of sentients), so to speak as opposed to individual sentients.
The <mother cell> highly vulnerable, can clear herself of sentients immediately.
Lense
Unit receptor
Recorder
Essential/qualitative sentients
An arc of a sentients. Receives and sends sound, picture, voice. (Sparcs, quarcs, arcs).

Mars 27th 05

Clusters in the line of stars.

M. Vygodsky, (A mathematical handbook #224)
Def: the derivative (Newton used fluxion) of a function is the limit to which tends the ratio of an infinitesimal increment in the function to a corresponding infinitesimal increment.
The derivative represents the rate of change of the function.
A. Gelfond proved, (in 1934), that logarithms were transcendental.
In logarithms naturalis the number e is irrational and transcendental.
Differential calculus.
1 finding the tangent of an arbitrary line
2 finding the velocity given an arbitrary law of motion.
The physico-mathematical sciences are either constructive or calculative.
Elementary geometry, rectilinear figures and the circle, their construction. Analytic geometry on the other hand is applied for the study of curves. The coordinate method. Calculations. Fermat, Descartes ,(plane curves). Euler,
(space curves and surfaces).
Eliminating, separating and squaring.
Vector: any directed segment and possesses thus direction, opposely, a scalar quantity does not possess one.
Scalar vector> products.
Triple product.
Standard>canonical
Collinearity>parallelism.
b
S y dx > b initial (the last) upper limit, a terminal (the ferst term) lower limit.
a
Triptometry.
Algebra of signs.

+*+
+*-
-*+
-*-
+*+
-*-
e
+
-
Real:
kinetic

restricted	rational
mechanical	potential

kE
mE
pE
Quitrits:

synchr	16
developm	27

W=p/t
T=w/p
P=w/t
(n=e+h)
2e+3h=2n+h
Logical or dialectical system.
Intuition acc to Hegel, (Philosophy of right 1820, #119).
′Etymologiccally, absicht (intention), implies abstraction, either the form of universality or the extraction of a particular aspect of the concrete thing′.

Kant, Immanuel, (Schriften zur Metaphysic und logik.
Wissenschaftliche buchgeschellschaft, Darmst. 1958
Originally pub 1770).

De mundi sensibilis atque intelligibilis forma et principis
Sectio 1, p. 18
Kant: *nam hic dissensus inter facultatem sensitivam et intellectualem (quam indolem mox exponam) nihil indigat, nisis, quas mens ab intellectu acceptas fert ideas abstracta, illas in concreto exsequi, et in intuitus connotare saepenumero non posse: daB di Erkenntniskraft die abgesonderten Vorstellungen, die sie vom Verstand erhalten hat, oftmals nicht in concreto ausführen und in Anschauungen umwandeln kann.*
Prolegomena: axiomen der anschauung
Antizipationen der Wahrnehmung *Analogien der Erfahrung*
Postulate ses emirischen Denkens überhaupt
>Dry rock
to read
Engels, (Principles, 1847 Pluto press Lndn ´71)
Proletariat: labor sale. No capital profit. Wholly dependent on demand for labor: the working class. Factory systems. Division of labour. The dominance of machinery; exclusive possession of all the means of subsistence. Specialisation. Corporative power.
Slaves
Proletarians
Serfs
7 Their liberation; the proletarian frees himself only by abolishing private property in general.
Christ as a social quanta.
Despise
Abuse
Deceive
Humiliate

Betray
Degrade
Fingers, muscles, fibres, back, lungs, stomach, face:
to feel ashamed.
Every muscular tone played at. Sentients made to interact.
A soul expresses shame through sentient>
muscular contraction> the very feeling sensed.
Corollary notions.
Betrayal of the old.
Forgotten locks.
Luminar units from the physical archive.
Units: compounds, properties of individual souls.
(Verified through the physical).
Binary tree.
Input, output, hidden node

April 7th 05

Taste, smell, touch.
Pain>pleasure.
Health>sickness.
Sight, hearing, understanding.
Feelings.
In embryo, double polarization begins to form.
As the individual develops, mixed feelings occur.
Their value is dependent on aforementioned polar opposites.
Mixed feelings in a fully-grown individual might
seem highly complex.
In fact they are not.
Of simple and mixed feelings
Love hate
Joy anger

. . . . . .
Trust distrust
Will unwill
Like dislike
Taste, smell, touch, sight and sound of liking.
The liking is determined by us. If not individually,
then by distraction, (taste of dislike, etc., attached).
Instinct: program, drive.
Visual
Verbal storage
Sentential
RNA
DNA
X (the mediant)
The organic soul.
Aminoacids, enzymes, molecules of feeling,
sensation, characteristic . . . hormones.
Scattered around the globe.
Substitutes: mechanical interpretation.
Loops.
Transmitters on playback.
Elan vitale moved to brain.
Liquid with defect.
Gazeuse with excess.
2mm tunnel in pupil, iris burnt.
The organizing principle.
Premises intertwined: a chance discovery.
Push and balance.
Eye-hand coordination.
Guard, nurture, cherish.
Arc and an anchor
Root, trunc, branches.

Representations.
Post-tongue; no sense of language.
Adorno, Negative dialectic
Abstraktion, retrospectiv, konkretion, authenzität. Negative dialektik ist kein antisystem: einheitsprinzip. Impuls. Intention des Zunächst. Zum heterogenen. Organon des Denkens.
Idee immanent.
Treue zur Stringens.
Denken ist Negieren, (p. 30). Pseudomorphose des Geistes.
Das Seiende.
Inborn strength.
Non conceptually mimetic:
enlivened beings.
Arena of faults
Top/bottom
The law of excluding middle. The synthetic find.
Radiation: Ångström
Max radiance: Planck
Black body radiance
Heat: strong, weak
Energy: much, little
Wavelength: long, short
Light particles measured in el-volts
E= hc/lambda
a= metres/Sec in the 2nd
pressure, density
1 metre: 1/300000 of a sec
Electronic mass 9,11*10 (in the -31) kg.

April 8th 05

No polar opposites. Mixed feelings recollected as the acoustics

of your muscular tones
Neurons. Signals control.
The soul penetrates the densest matter.
The strings made by burning light, the flames of God, if you like, can be seen on a nano-scale. Every individual being is thus carved since embryo.
Genetic anomaly?
Evil has to be coped with. No other apparent way.
Too much energy to hold them captives.
The art of carving.
The brain carves itself in a way. Energetic people have a much wider spectrum.
But we intervene whenever we think is necessary.
Currents from the axiom of intuition. And polar opposites.
You can bend the line up to a point. That´s the relativity.
Main currents.
Anger and energy release. Metabolism.
A distinct molecular structure.
The energy that can penetrate walls.
Barriers: some have to be monitored, as we care for different results at times.
To trigger energy release.
Lower energy, other currents. Differing kinds.
Constant flow. By sleep the flow changes.
System receptors.
Information-processing systems.
Neurochemical changes.
Axiom of sensory inputs.
Store-retrieve.
Shaped by evolution..
Design.
Heredity: inherited features.

Causal mechanism.
Negative feedback.
Genes: units of heredity.
Cognitive, motivational and emotional processes.
Mediating activities.
Proteins are specific sequences of amino acids. Proteins act either as structural molecules in cells or as enzymes that catalyze chemical reactions.
Produced by genes which control the synthesis.
Inability to convert amino acids: brain damage.
Conversion process.
Alleles. Each allele codes proteins.
(Recessive: both dominant).
Changes in the DNA code occur by radiation, or
by chemical agents.
By intervention, natural selection. To eliminate harmful characteristics
Probing areas in the brain.
Nature´s way to cope with pain: reorganization to receive input from elsewhere.
Stray electronic signals.
A pulsar: pulses of energy.
Air.
When air comes short, chain reaction might occur.
Decline, decay, degrade.
Motion: more velocity. Higher altitude. More penetrating power.
Ionization.
Lose energy without gain.
Diplomacy: from a mountain to a mountain.
Co-conducting.
Super-conductors.
Atmospheric layers.

The technique

Developed in the 60ies. Discovered by a famous doctor.

The Technique has been evolving enormously for the past 45 years. We have made some experiments on you. Some have failed. Others have done wonders.

The process of draining your glanular receptors in your head is centrifugal.

A part of the soul´s RNA chain is taken off and put in the entelect. That part copies everything the soul does. To infiltrate the unit/ sentient, (that distinct RNA part), with the use of your eyes!

Units connected to musculature system. When it´s owner does something, the unit ´imitates´. Proof of deed.

The unit, the part of the original RNA chain, that was taken off can in itself not move.

Because of units their work might seem easier as the individual soul can give the same sentient in more than one and more than two other bodies. Some, I wish to think might possess multiple units. The copious unit cannot be used to hide a bad motive.

The masse populaire then controlled, or at least semi-conducted.

Benefit: no physical entity is unattended.

: eases work

: unit in entelect will put bad deeds at odds.

Feelings-electricity.

Chemical bounds.

From units to units.

It circulates

Kinetic movement

Securities froed

Give false warnings, lie about profit, misleading info . . .

Spirits and business ethics.

A&B I B&C

A&B
A1
B2
C3
B&C

______

A&C3
Legal?
Firms A, B, C
A&B
A+
B-
A sells to B
B value goes up
Cn Buys in B
Value goes to the rooftop
A Sells
The mechanic nightingale.
We form your soul. It ages, that´s it.
Orange network.

Mai 10th 05

New scientist, (Aug 18, 1990).
“The vibration spectrum of a diamond.
Waves of light and sound.
Splitting up (ordinary) light with a prism to obtain a spectrum of colors.
The distinction between sound waves and light-waves disputed.
Can be ´seen´, or ´heard´.
Zero gravity> weightless condition”.
New Scientist, (Jan-90)

"Monitoring radioactivity
Sharing electric charge.
VLF very low frequency radio waves.
Reflected.
Refracted.
Attenuated.
Depending on the ionization of the ionosphere:
a charged layer"
Chemistry.
Identify the patterns of elements.
Rings.
Chains.

Mai 11th 05

Air	Gas
Fluid	Liquid
An organism with a distinct molecular mass,
fluid and air.
An explosive lethal pressure, fervent heat
Void (astronauts), the body perishing at sea, no upheaval. The soul perishes.
An individual needs at the hour of departure to become whole.
The soul is in a way constructed at the hour of dep. Midwifes.
No matter how structurally stable molecules may be.
Suicidal sacrifice.
Knowledge is, as ever so often stated, a dangerous weapon.
To hide and act as another from within.
Gravitation/ nuclear force/interactions and radioactivity.
The moments before radioactivity: fundamental attractions.
Sound.

Bound by light.
Ultrasound.
Infrasound.
Waves of light.
Kernel-strings.
Kernel-particles.
Waves and sound.
vibration
balance
frequency
strength/volume
distance
Sound fades away.
Light gets absorbed.
Hindrances
Excitement   disappointment
    Different motives, different points in shape space.
The lobby.
Multiple screens.
Processors.
Communication.
    Linked.
    To make and receive words or images; to sense.
More liberal.
    Your own fixed line.
    Direct reading group etc.
Digital conversion,
sharing facilities
Gliding carriers,
conducting currents.
Penetrate, scan.
A supervision: the agency

Two nations using the same bases of technology, yet incompatible
The lase.
Lasing the physical.
Voidal core.
Ions from gases.
The pleasing understanding.
The pains of not understanding,
neither right nor wrong
The ratio between void and space.
The very order of being at stake. Good and bad. Our wrongdoers.
Images in fluid.
The light of movement is so violent because we opened the pupil onto the iris.
Hermaphrodite-program
Siva
Needle-eyes
Standard procedure during tomie:
preparation,
made to lie down,
paralyzed. Minimal physical activity:

Glands
Fibres
Brain
Muscles
Eyes
Etc., hantered with

If serious, backup team. Signals interior. Someone helps you breath if necessary.
Your brain. Your muscular atonement
cut to the outmost.

All the horribles, all the terribles.
The limit.
Your nose, mouth, ears
sort of cut off.
Regional shortcut
The retina. Your eyes respectively had to be made gazeuse again.
Mining the physical: gas. Motoric system. Movement.

June 17th 05

After an interval.
Eyes 20 m
20 cm
The family.
Droplets.
Right ear. System down. Only the frontal functioning.

The plains of heaven.
Penetrating the ancle. Imps into the blood.

June 24th 05

Eyes. Finesse.
Sentients
Imps
Arc
Optic tubes, neural > filtres
fibre cells
chemistry
State: compound
mass

pressure
work
power
energy
Stable, unstable
Attract, distract
Cyclic. Product, movement, charges.
Physical sentential; intermediaere.
Optic reading. The imp as a sentential product.
The past infiltrated.
A sentential image from the physical read optically.
Physical proof. Trigger of movement. Imitates the act.
Illumines the deed:
coloric agitation
Catalogue of imps.
Imitating physical movement.

June 28th 05

Adjustment.
The midbrain.
The basest solution.
Embryo carving.
The escapists eye within.

July 1st 05

The most beauteous flower with the deadliest smell.
Move grow.
Pneuma.
Dominant Retrieving
Product
Extracting colors.

Move grow.
Taste smell.
Freshness.
Capacity: energy. Division.
Labo drugs.
Shape space.
Antimatter by convenience.
As a fit.
Radiant light.
Elimination.
Decay.
Ruin.
Colliding energy emitting . . . ?

Norm.

Real memory from above
Physical memory, reflex: a place you have been to before.

Solar memory: feelings.
Environmental.
Learn.

Joint evolution. Any measure off point. No axiom to build upon.

Physical response
might include
basic learning from experience,
′recognition′, ′recollection′, of a place, faces . . .
a language of some kind
instrumental skills

Solar energy, feelings, sense
Structural evolution.
New molecules.
Recreation.
Ace and base and . . .

Charged differently.
Photomatic.
Sheltering eye.
Light movement.

July 20th 05

Glanular points
in body and in brain
Signs of movement
Red Blue
Beam
Ray
Radiance, gleaming.
Light, air: form
Time.
Speed.
Void. Space.
Optic measure.
A scale.
From color< (maximized in scale).
La source, l´arc,
vide elastique.
Dew.
Clear.
Color sensitive.
Medium: bridge. Passover.
Brightness: volume, strength.
Kernel axiom.
An inverse space: a light of movement, motion:
emotional void.
Our units. Your pigment.

Deconstruct: chaos > change > construction: order
Interior design. Motion dry.
Filament of detection   Interaction
Infiltrate   With consonance
Clock-watch.
Verbal groups, (biological).
Nominal threads, (organic).
Threads
Axiom
Measure
Negation
Unit
h
in/out
l
hit, strike
To tame the hide
Light   dense
Anc of now, flux, eternity.
A ring, circle.
instrument vehementer
aer of age: act in time
Soft   hard
Void of aer,
thought unwanted.
A vivid thing. A signalling
kind. Moving beams.
Beams and wires
Lay
Wide dimensional
a++
i--

e--
Ionic bound
Movement. Force increased. Acceleration. Shape space product.
A neutralized
1 waste
2 no mass, (virtually).
Work: increase, decrease. Move.
0 (has)
0+ (gives)
0- (takes)
Movement, motion > numerical, by measure.
Force to overcome hindrances
system
Dynamic material
person
Scientific, character, social, mathematic.
Gravity.
Motion as opposed to movement.
Bad thought that causes movement.
+
Balanced number
-
stable. Static.
Charged with number. Loaded,
magnetic
Attract, distract, stabilize.
Norm. Standard procedure.
Veritably out of it.
Extreme=outmost limit.
(Innermostly conservative)
-(Couldn´t tame the light longer).

Electric number
Element
Atom
        Melting
        Boiling   point
        Freezing
Subatomic structure
Nucleus
Nuclear
Nucleotide
Electric
Particle   spin
Isotopic
Gyric, panagyric
                    -(Never falter).
Muscular contraction/connection.
Fibres from tissue.
Glanular fluid in tissues.
Tubes and stocks.
        Archives.
Fusion
Fission
*
Gravitational field.
Luminar shield.
Radiophonal waves
        waves of light
        optic particles.
Infinitesmal calculus.
Strength constrained.
Ceramic,
textile

Voidal creation.
Amplified reasoning.
A prophecy made to fit.
The light, light, light
A lie to beat a sublime fiend.
The magnified awoke the mighty.
The daring in the now of Rock, the choral tune.
Such horrendous aquarelle of lies and deceit.

July 21st 05

Prayer of light:
forgive us our sins against the flesh.
Med.rep.
Tomie.
Intestines, liver, stomach and such,
circulating fluid, (150 per min)
heart stopped twice.
Archive from wherever.
Recr: fewer loops.
To verify whether women give themselves freely
or are actually violated.
To bring into light drug abuse.
Whether the heart is attacked.
Droplets (nodes) irritate. Elan in droplets.
Carriers from overseas.
24 fluids in place
controlled by air
strong as light.
Prototypes of movement.
We get what the insects leave us.
elan is see.

Edit.
When edited, one loses, in proportion to a crime, capacity. A part of his vital is simply removed.

July 25th 05

More cuts. Through organs, peniscular, heart conn, other routes. Carvature in the brain. Ratio: to simulate movement, to recreate the deed, to tame the light
Fluid from glands. Places found. Droplets made.
Light through void.
Abdominal man.
Women of the world have been hiding their sexual behaviour.
The organ. Console.
“sound from pipes set on wind chests
the flow of air controlled by mechanisms
keyboards”
Panpipe of some sort.
Hydraulis, 3d century bc.
Levers admit or shut of the wind.
Opposing truth, action committed breaks through as crime.

July 26th 05

Axon: signals neural response
Tunnelled through the cytoplasma of millions of cells in the organs. Might ruin the micro body and or the mitochondria in the cells.
A new technique.
The art of crime. Where to hide their sick emotions, longings.
Mouth, nose.
Constant pressure in head

light through testicles.
Two genders.
Organic layout.
Motoric plan.
Lose energy, gain time.
Adjustment archive.
Color proof.
Verified immediately.
D red
D blue present in brain
Cr. green
Mixed colors
Feelings
Sense: smell
Spine: intact
heart
regions
abdomen
Physical points in shape space
found by penetration:
when point in void is known.
Point:
pure
shape
kernel
Escalation
theory
. model
within the Godhead.
Total info awareness.
Enabled danger group,
able danger.

Tear or cover,
down or up
Sub-agenda.
Commition.
D.E.M.
Space faring.
Take deep.
Athenic/palladial spots.
Society is changing fast.
Carriers in every chamber now, uniting the see.

Aug 1st 05

Med.rep.
Cut through back and breast, before from back onto spine.
Lungs out.
Modification of individuals- to supervise.
Concealment.
Skeletal key. Bones bent in the void to simulate
Recreate movement. Pouring sentients in to the points in
Your organs. Heart stopped for an instant to change
from one region to another.
Imps from around the globe.
Pour into the eyes, scanned through
glanular fluid. Crooks found.
Anchor themselves to the physical.
Lasing down upon every living being.
Bliss: silvery flash.
:people lose desired memory.
Wrongdoers trapped and bound in the physical.
Give to those who have earned it.
The physical as a devil´s trap

Love reasoning: spine fluid and yellow.
Color search.
The tiniest speck, , primary or mixed, sometimes flag-like, tricolor: the proof we need for deed committed. Simple ain´t it?
Circulation of fluid: kinetic energy.

Aug 21st 05

new route through R. kidney.
Other cuts from abdominal.
Eyes. More burns. Lost depth.
searching points.
nodes, the muscular/organic analogy.
The woes.
Epidemi
Famine
Environment
    Atmospheric collapse
    Chemical poisoning
War
    Oppression, racial, ethnic . . .
Pollution
Technological failure . . .

Sept 02nd 025

The chemical. The hurt.
Focus on the moon.
A message.
Evil staircase.
A spiral to unwind.
Attacks from wherever.

Bad light.
Leftovers.
Remnants in time.
The deeds onto the physical.
An erring mass.
The basis from which people feels. The glanular product.

Sept 2nd 05

In ′drop of life' : nerves cut from body.
Butt
Breast
Knees
Elbows
Nipples
The soles
Different wavelengths, types of beams.
Automaticity.
Overcharge.
Over amplified .
Lased down upon.
Under the gun.
Stove/heater.
Joint cut.
The reservoir.
Placement of metals.
Creating compounds:
take in charge at low voltage, work>high voltage.
Currents all through your body.
Before your eyes, -or- voidal movement.
Coil up in the pearly.
Energy pr charge.

AC
DC
X
Current around a closed circuit.
.volume – flow – voltage.
Metalloids.
Ground wire.
    Current throught.
Conductor.
    Protection against fault, chock.
    Spending charges.
    Atmosphere reservoir.
Laser action.
Scaling with the wavelength.
Lasers as step-down transformers.
Discharging to earth.
Electrical discharge energy transfer by collision.
Operating from the infrared and from the optical.
Transported cross the Atlantic.
Superfluid. Condensed atoms.
Very high heat conductivity.
Solubility of gases.
Scaling inactives through the gas chamber.
Turn active at will in the physical.
Transports: pass it through.
    Satellites, mirrors, (individual).
    The lites.
Containers.
Storage.
        Spreading beams.
Increasing the physical volume.
Cavitation: elasticity, volume.

Grab, capture.
Diverging lense.
Holography: not a focused image. Light is reflected and combined with a reference beam. Parallax. Impact: more info.
Holograms in midair.
Images, holograms.
Depth, parallax, continuity, scalable
The image and the act.
The particular included. The entire.
Pore.
Cylinder.
Cerebrospinal fluid
Teutorium cerebelli
(int. organs)
Cut from the arterey
pineal
Controls: pituitary: eyes
thalamus: ears
The magazine from the controls back to the calcari.
Through the orbital tract.
The parietal branch (the zenith) the auriole of the vault.
Distinct nerves disconnected momentarily.
New tracts.
Maya-tracts and nodes. To create false movements, (to find the see).
Skeletal split between.
Ulna and radius.
Tibia and fibula.
False joints.
Epithelial: epidermis, skin, linings.
Cells>tissues: connective. Dermis, bone and fat tissue.
Muscle.

Nerve.
Trigger movement.
a)directly, e.g. electrical signalling
b) by metabolism, (heating glands . . .).
Refraction of thoughts.
Bifurcation of words.
Scathing incidence.
Refract or reverberate,
Balance,
fall apart.
De/capsulated.
Counterpart.
Diminish.
+
0
1
Waves.
Near micro.
Far infrared.
Thermal.
Med.rep
Incr. bloodstream.
Inf your heart.
<Your heartabeat is out a year ago>.
Tomie:
2 hours later pouring in
The calcary heated to cause pressure> to agitate.
>On a silver plate remember.
A year ago the brain nerves were being cut from the cranium. Just now the last nerves from the calcary; the calcary and the skin.
The cranium, the spleen, (a hand fx).

Or in the same manner.
The cranium and the cerebral tissue,
the brain itself.
The void. Movement of light as a ratio of the physical cuts,
The cuts and carvings on the other hand made to replicate
Distinct real movement, as was said.
The calcary, some organs, parts of the lower back.
Skin. Epidermis, tissues. The nodes.
=Deinde unde,(inside out).
Asymetric: left side.
Horizontal circulation again.
Infused by heating.
Thermodynamic energy..

Sept 5th 05

A hierarchy within a hierarchy came apart.
Effect. Butterfly: suicidals, freak accidents, coincidences
when the point in space vanishes.
Pouring in:
He rain
Na rain
Electromagnetic.
nserted manually,
(other places).
Mainzer: the graphic notation from Brown, C, 1864:
Orientation of atoms.
Kekulé: Quadrivalent carbon atoms.
A structural formula.
Charge density p(r)
R= space vector of an electron.
Nuclei attract.

Linear molecules.
Such as:
N=N nitrogen
C=O carbon monoxide
Axis of rotation
Plane of reflection.
Vertical/horizontal.
Equilibrium state of molecules.
Rotational motions.
Sigma bonds n or h = plane of reflection.
Chirality.
l or r handed receptors.
Cn=axis of rotation.
Parity.
Strong, (restrained).
Weak forces, (eased).
Rotary reflection.
Axis of reflection.
Hybridation of the orbitals.
Point group.

Sept 9th 05

Transatlantic lock.
Chemistry.
Smells.
Sedative.
Movement of the electrical current
between hemispheres, (regions).
Around lips.
Voidal dimension because of cuts in cranium.
Coil.

Turbine.
I individual sentients
II (make them think)
III send it.
Finishing your colors.
Typing elements.
The driest ray.
Radiant.
Absorbing.
Strings. Tubes.
1 Organs first,
2 through and around the dermis, bones, muscles, most of the nodes,
3 air.
There is excess light in the world.
To be and to appear.
Without really being there.
i or f : initial, final.
Lased through to the core.
Pinpointed from up above.
Red lightning.
Lame for a while.
Fluid in marrow.
Left metatarsa, toes.
Insert our organic waste.
Sedative.
Cuts and carvings.
Right hand. Tube inserted.
R. hemisphere. Insert fluid.
Asymmetric.
Zenith.
Constant ear.

Head off shoulders, (weeks ago).
Tubes.
R. hemi. Cuts right through.
Pineal.
Carving the thoracic area,
The heart.
No hold for the see.
Movement visualised.

Sept 15th 05

Those who do a bad deed by heart.
Voidal point.
Transform.
The ventriloquist and the puppeteer.
Spiralling up the spine.
Cerebelli: lower back, bit of the spine.
Obituary, sense, hearing.
Thalamus: heart, sense muscles, touch, taste.
Pineal: sex organs.
Radiometric.
Day by day.
Scatter.
Vertical stray.
Sup vis <10mm
Inf vis 400nm-800nm.
Expanding the spectrum.
Absorb energy, emit.
Radioactive or electromagnetic.
Electric permittivity.
Magnetic permeability.
Stability:

mass
force
charge
Transition state.
    Vo
o-   o+
:each flavor, 3 colors
R+b+y=Wh
e-   e--
-e   --e
Collision.
Decay.
Transition.
Event shows energy, lack of momentum.
Dissociation or fission.
Gluon: strong force betw. Nucleons.
Photon: electromagnetic, (activity).
W&Z: weak interactions..
Gravitational interactions.
Influence of charges: interactive force, electric field, voltage.
Stretch out through the sky
span the world.

Sept 29th 05

Veritably through regions.
Gain: less pain, cascade of sinners.
Loss: no feel.
The utilitarian foundation.
Panopticon.
Fabric of felicity.
:We took you over the threshold of pain.

Dust, stain.
Stealing air.
1 Enter the physical. A chemical agent: the solvant. He, Na.
2 Specks, stardust, quantized compound, radioactive elements, (F, Ar, Ni).
3 The works.
4 Exit.
Blood heated, glands.
Energy stolen.
Hot or cold for you.
Absolute negativity.
Corroding.
Decaying.
Guard.
Search.
Clean, (recycling light).
Pouring through the nose. Nasal cavity.
To the center of the brain.
Entelects: not likened. Hated by a definite percentage. Capsule of espionage.
Too much work in and around them. Arouse shame, pity. To be able to work, people has to be hardened.

Sept 29th, 05

Layer 11th.
Tube in fake vagina.
Elements in blood plasma.
Compounds of no use for the dead.
A burden in fact.
Spiralling inwards.
Just to hold it back.

We now pore it back to you.
To decompose the product consumed by us.
The chemical is toxic.
The waste eats you up.
It´s a by-product of an activity
-how we use light.
The compound is charged
-can cause physical pain if released.
It causes us to act in a distinct manner.
Is sort of infects thought.
Victims of bad thought.
Superior just as inferior.
From the sin to the very crime.
Thought materializing –cause of action.

Sept 30th 05

15hr: cosmoclock: now we died a little.
Irrigating the physical.
Scalpel of luminous vigour.
Dissecting.
(From grace) portal opened.
Calculating the end of the world.
Cleaning atmosphere.
We´re constructing:
some kind of a tunnel.
For you to bear.
Purifying souls.
-(*On vous sent mourir*).
Light.
Frequencies, bandwith.
Atmospheric.

Famine in Africa:
all you can do is talk about it.
Animal soul.
          Pneumatic.
    Solar.
Instinct.
Hurl.
Compounds from the atmosphere.
Glued onto you, as magnet attracts.
Also: constructed to ease work; more automatic.
Just pore it down, through the cone, into the cavity:
chemical man.
Cavity betw. Brain and skull, as big as
the palms of the hands, for chemicals.
Crystal blue specks all around the body,
antennas, devices.
Diplomatics: where mountains meet; in the troposphere.
=Imps in body. Nodal coloric. Effectively
in brain, through eyes resp.
Neck.
Left eye hurts: the vault
Heart, genitals.
Straying through face, eyes.
Real face out..
Portal density.
Light of every known spectrum pouring,
really threshing itself in and through.
Monitored.
Crystaloid.
Crystal rays.
Running the stock through the lock.
From a region through a region.

From the epidermis,
in every river, to the marrow.
Every organ.
The chemistry makes one act
-in a way.
Straying by numbers.
Thought control.
Stronger than any a see.
One more line on top of a layer, through the skelette,
jawbone, cut to the outmost. Cheek. Socket.
Point objective:
white, coldest ray.
Cavity created.
E!=entelect.
Mining the physical.
Chemicals>E!>weapons
or
=Concept
Beta ray 30x
Gamma ray 1x >E!
Circulation
Heat concept
Some equatic force

Okt 10th 05

War in the stratosphere.
You´re the core of it.
Nest of nests found.
We´re literally throwing weapons up from you.
As a part of a propaganda machine.
Enter list of numbers.

Your nation lost independence.
Youth cut off from the elder.
Obedience, the master concept.
Absolutism, paternalism, consent.
A cap might come in handy for your nation.
He & some elements/compounds in the bloodstream.
Compressed compounds in the gazeuse system.
Compounds from the physical.
Work done.
Force used.
Energy released.
Product concept
Metabolism.
Collimating.
Nuance in head along with hammering bloodstream.
Strays from above or below, from the front or back; through sockets, ears, in between cheekbones, through the skull.
Components separated from the bloodstream.
Deterrence >retribution/correction.
False consciousness.
Workforce controlled, workers demand diminished.
The production of armament.
Warfare state.
Composing, deconstructing the counter-culture:
world government.
The nasal cavity: to pore light.
Ginsberg:
the soul is a whore.
Moloch & wargods,
Birdbrain,
Mother-gas,
Ginsberg:

Jesus Christ was killed by his mob.
Hired soldiers did the job.
Highly radioactive.
Light scattering from your head.
The pressure we build in your head,
the chemicals we put in.
The ionization we need to develop.
The light we emit in you
comes from a strange place.
Constantly pecking the lobe.
Elemental tunnel.
To you, who did not lose your fatherland.
Milton:
embryon atoms.
The dark materials.
Aerial weaponry.
Nuclear.
Destroying the very essence.
Blasting adversaries.
The numbers they lost after the Twin towers attack.
A scelette dressed in flags.
          Lasellite
Instrumented   sacramented
Used to torture and kill.
Building up an arsenal.
Armistice.
Mining anew.
The adversary
smuggling weapons to the states.
Defend against electronic attacks.
Space tracking and communications network.
Tracks and controls more than 50 satellites at a time.

Satellite.
Digital camera and image sensors.
Strar trackers and reaction wheels.
Antennas and transmitter/receiver.
Flight computer.
Pointing control-propulsion mechanism or momentum wheels.

A
B B
A C A
A

A A
A B
B A
C
A B
B A
A A

Short of protein.
Gas put in lungs.
Adjusted by heating.
Add:
threefold portions.
From any a ordinary.
Retinas compressed with gas.
Regions in the brain also.
Itch along the veins.
Alienation.

Oct 21st 05

Adjustment: heart made to stop for a fraction.

Face off.
Lungs off.
Stomack.
Penetrate to a point in back, genitals.
Hands, feet.
Specks inserted. In the node of hearts.
Something within you came to ruins
adrenal needs to be helped once in a while.
Eyes: left: cuts; the whiteness.
We died from you.
We´re not with you as before.
Alpha decay:
Ejection of a helium nucleus,
(2 protons, 2 neutrons)
>gamma radiation.
Beta decay.
Ejection of an electron
from a neutron
>gamma radiation
Electron capture.
Head.
Heart.
Abdomen .
Genitals.
The head as the body as a whole.
Hardened as a happy dead.
Heated 60°.
Abdomen.
Hands recreated
>wide dimension.
Cuts, regions turned on and off.
Bonal cuts.

Face, nose
L.ear ultra “eye”.
Lumens
Left hemi, right.
Slid right through.
Hands on or off.
Feet, genitals, organs.
Organs dead.
Gain: heart does not
beat in its proper region ratio on body
N in the power of 8.
Node of heart.
Still activating if you will.
Your mother used as a decoy.
Shot in heart.
Marked in the physical.
Verified:
The deadlies.
Eyes hurt a bit.
Irritation in the skin,
(toxic chemical and the light constantly ′fondling′).
Vertigo, dizziness.
Heart slightly off beat.
Constant headache.
Due to a) cuts in epidermis, the skull, the chemicals in the cavity
Between the skull and brain, b) heating, c) movement of light, other.
The young:
shortage of material.
Movement,
Sight comes short.

Social and moral implements.
Nerves,
(heart)beat
Poisons.
Burnt flesh.
Wood.
Crooked round.
Two flags out
>diplomacy.
Assigned, magnified souls.
Chemical structure.
Eyes. Tubes prolonged.
Abundance of essence.
Your fluid, a bit of your essence and essence from us.
Bonal structure decomposed.
Rotates at will now.
Spleen opened.
U.V. light.
Coloring specks.
Nuance.
Specks in cavity.
He, __, in blood.
Ultra light.
Charge.
Magnetism.
Ionization.
Electric.
Deeds illumined.

Okt 25th 05

Boiling point error.

Photonic.
Tele/scopic.
Hardwired, grinded.
One conscientious mind.
Wide spectrum.
Nodes of movement.
“A net to catch the wind”.
The structure within..
Givenness: Withness of the body.
Phododynamic.
  Bird on a wire.
Fluorescent, photonic.
Nanomesh automated.
Short circuit.
El-shock.
Ball joint.

Nov 20th 05

Ionization tables.
The here and now illusive.
The being as such
synchronized, controlled impersonally.
Mediated through light.
We can see, (in you) ,
any point blank place in the spheres.
The clinic.
Defects.
Reverbrance.
To create a void in the atmosphere.
Cause: chemical chain-reaction.
Pulsative attraction.

Strain.
A certain figure: a definite number.
Animated response.
Vibrating resonance.
Equipments for eyes.
Closed cycle.
  Space group transformation.
  Change in electrical properties.
Scattering power.
Network of compounds.
Different temperature, pressure: different structure.
Trigger an irreversible mechanical behaviour..

Nov 4th 05

A nightingale´s shiver.
The heart. Has taken some time. Parted from
the lower part of the body. From the upper part.
Two cuts up the spine. The spine had been
somewhat hantered with, nerves, fluid . . .
Some genuine device had to be invented
for this grand scheme.
Cuts through some centrifugal areas,
veins, nerves, tubes from the heart, through it,
(penetrated).
Bridges apart
Two short cuts in the brain, (ratio).
The godly ventriculation system turned back on, aerial chemistry.
Turkis, Dusky pink. This summer for women.
This morning. Purple. Stomach.
The nodes of the heart.
The mind their extension.

Two weeks ago.
-(I found the wind).
Pneumatic with light particles.
Scattering back and forth.
Hurting living beings.
Blazing, sweeping, chasing.
A week ago.
The densest mesh.
200+ sticks stand out of your face.
A muscular net inserted at will.
Defects.
Vanity fought..
Panopticon filled up by you.
Structure.
Net through.
Gaseuze tunnels.
´Eyes´, oculars for those in charge.
Net over and beyond.
Put on or taken off at will,
(wide spectrum).
Been trying to organize the loops, feedbacks,
as to what movement, sight triggers what.
Easier for us. Highly mechanic
The dermis ripped off..
Under the cloud as usual.
Attract active specs, inactivated, (so to speak).
In the physical. No ionisation.
He, _, in ventriculation system
oxygen from the corpse.
Lesser than the least
Walk the street.
The sky is the fist.

Nov 10th 05

Nuance.
Beating vibration.
Units in place.
Viscosity
Electrogen,
halogen,
metalloid.
Undular symmetric.
Muscular (organic) harmonic.
Color reading
Spectral, (electrogenic).
Electric-sound.
Wide spectrum A, Rn.
Spheric.
Special effects..
Partly autonomous.
Eases work.
Carrier unit.
Light monitors
Sound amplifiers
Trigger reverbrance, loops.
Tools
Unit
Equipment
Gear
Instrument
Props
Structures
Lattices
Modified socially

Light improvement. Too bold an attempt.
Had to crush down.

Nov 11th 05

New colors, (discount).
Inserted with fluid and some
gas mixture in the godly ventriculations system.
New device.
Purifying air, both by heating or cooling,
with pressure and electricity.
Spinal fluid, adrenalin, stomach.
Cultural virtues
Discipline
Respect
Dignity
  And like
Emerald
  Puissance

Nov 16th 05

50 new colors.
Purifying images.
Cured by color and light.
Abundance of colors.
Leaves place for experiments.
Mix them differently.
Another cavitation.
Yr eyes are somewhat different now.
Work as usual.
  Nuance

Panopticon
Corrections
In short: helping people
Volume:
sound penetrating.
Sound and second sound barrier.
Vision. Another bandwith.
Vibration
Frequency
Before and after is now.
Anchor in bone.
The pouring eye.
The body as a net.
A million faces.
All in one.
Skin peeled off.
Cuts in and around heart.
Voidal lines have changed.
More cuts around, within.
Sex organs, abdomen.

Nov 19th 05

Light anchored, grinded.
Vision and width.
Active>magnetic>inactive.
Spectacles.
Charlie fine
Point, point
dot and line
here is Charlie fine
the neck is long

his stomach strong
hands, hands
see how he stands
hair, hair, hair,
here is Charlie fine
Hands now we him lend
fingers on each hand
If you give him a hat
you know where he´s at
Farao´s dog.
A plaything for his children.
A point in the
line of stars.
Nuance, specs.
Embryo.
Supposed skin disease.
How we die:
  Ignorant,
  out of fright,
  bad memory and such.
Red light.
O, He, Rn,
Irritating the skin: epidermis.
Charge attack,
(lightning green).
Gaysers in brain.
Shot from plantar.
Marrow.
More viscocity.
Cavity.
The tender touch.
Band. A feet per sec.

Band of sight. Burning at the corner of the retina.
Your kinetic.
  Oxygen, (leftovers from blood).
  Proteins.
  The pulsative.
  Muscles.
Fluid with light red.
Viscosity.
Tomies.
Fluid+ chemicals.
Cuts and carves.
Virility – amino acids.
Force field.
Knees. Elbows, (due to cuts through joints).
The back of head, (raw use only).
Embracing wind.
As one body.
Illusive might and cruelty.
Enter fluid in circulation godly.
The center of the brain.
Cultivation.
B>R
B/R>abc
Hardly a light.
  An instrument.
  A reflection.
Phenal acid.
Right eye nanotech.
The bands.
From a color to a speck.
Seat empty.
The three glands in the brain. Out.

Pressing regions.
Golem point.
Color boreal.
Fx bipyramidal. Closed. Open.

Nov 29th 05

Flashbound. Bliss.
Exuberance.
Exhilarating.
Adhesive.
Sound device.
Meteoric.
The cosmic route changed: tunnelled through you.
Sight improvement.
(Good and bad) light.
quantizing
L > D
destructing
The 15th ? network down. 2nd time.
A globe.
Ratio of the lot.
Bad/wrong move.
Absorbing now.
The repelling force.
Can´t bind to the flesh.
Scelette bound.
Grinded
El currents go through you.
El shock.
a) To punish
b) To change individual thought.

thought in a way inactive.
Image/word.
Action.
Movement.
Motion.
Uniform inactiveness.
The might to activate.
To change a thought.
Under the silver lining.
Verifying, correcting.
Quants of red and blue.
Frequencies.
Showers.
A mammoth thing to do.
As old as mankind.
Sth all of us did,
some of us do
none of us will.
How we touch the body.
Compound Unit/structure
Circular relativity.
Threshed upon, made inactive.
Peculiarities of a forgotten past.
Habit bound; released,
(numeric traces, chemical warfare.
Sure to lie upon and hide if you can).
Lost control.
Can´t trace back.
Memories unfold.
Mammoth species.
The cure in you.
The pains before, the moment of release; relief.

More than bad deeds now, as you know.
Erroneous past corrected.
Historical vision established.

Des 1st 05

New elans.
Gathering clouds.
  Bound to be released.
Erroneous task, (some say).
Too many mistakes.
Thought just drifts.
Casualties.
Innocent victims.
Then realize the gain.
We started something.
And we will succeed.
Summoned upon you.
Introduced by electricity.
A superimposition.
The very birth.
The direction of life.
Correcting wrong deeds.
For bad: imprisonment, edit, death.
All through you in a way.
An arcane pulse.
Abdo-net out, instrument out of head.
Peniscular cut, instrument inserted.
Figures. A net of eyes to trap.
When you were out:
made to enter one by one.
Now. Hundreds at a time.

Transatlantic.
Seven seas.
Disorder or disease.
Light sewing in fluid.
Electromagnetic.
Body parts: fluid.
Through hands,
socket, nose, nerve ends.
Along circulation
into spine and up.
Saliva.
In head.
Engine.
(A turbine), cables.
The end of the world:
the atmosphere will collapse.
Light attacking.
A structure came tumbling down.
The emperor´s nightingale.
Currency: points in air space.
                  work
To change a world. Mediaeval transparency. The symptom.
Accuracy. Nomad transport. Beneficiaries gone astray.
Thorough investigation. Help is on the way.
What came tumbling must come up.
The French don´t like it any more than we do.
All of Asia on the verge.
An island enchanted.
Structural failure in the States.
The Commonwealth has turned direction.
Officers running all throughout Europe.
African youth, (the sunscreen).

Arabic network does not work properly.
Highly politic. The remedy; expenses.
World council. One globe. Save and be saved.
Eastern Europe still holds back..
Due to the defense mechanism.
Light by types.
-(Be kind to us now).
Stop or be stopped.
-(It´s crowded in here).
Blazing light.
Point physical: you.
Will it break?
More than a structure, then.
Trigonal.
Cubic.
Binary cause.
We hurt people. That must end.
Eases work.
The ultimate bend.

Des 03 05

Emerald spectrum.
Not so much turkis and violet now.
Tanning.
Transparence.
Quantic device.
Cavity.
Up and through.
Uterus.
Embryo.
Enter: shades of white

- transparent
- emerald

Leuchaphelos took the blow,
(the white-headed Eagle. The U.S. respectively).
The might to convince.
Blazing on earth.
Something happened to you and your father
Simultaneously.
What you say is what you do.
Better think twice.
Mining again.
Propeller.
Devices.
A blast in aerial space.
Armory.
-(You´re my archibald).
The flames of their gray silver
the sparks of their old steel.
Famine in the world:
woes of neighbouring nations

Des 5th 05

Med rep.
Color adjustment: respiration.
´Come´.
Chemical shower: toxic, saliva, taste, raining specs.
Vires and nets in genitals, abdomen.
Well equipped as usual.
When the steel breaks.
A panda and a bear.
Devices within attract and repel compounds.

The energy needed.
A kind of an aftermath.
Counterattack.
Quite a few great men lost.
The spheres over east Asia bleeding.
Figures. In the power of five.
-(Within the warhead, they say now).
You hold the eye for us.
Solar equipment.
A network of Entelects.
Muslims enslaved, as it were,
in the physical.
Memory, thought, movement controlled at will.
Back to the construction site, then.
The Russians and the Chinese oppose strongly.
Egyptians closed their vault.
Italia: lost connection.
The Germans look the other way.
The Japanese stand aside.
They have to be supervised. We don´t want any
bloodied attacks, as we have witnessed in ten, twelve countries
for the last four years. (America, Israel, Spain, Indonesia,
Turkie, Russia, Egypt, England . . .).
The disarmament in progress.
Here is the see, see
nowhere to flee, flee.
Captives amended
something has ended.
Light.
Density
w>y>b>r

Des 7th 05

Inside your back.
Armpit. Nerve ends on the sides.
Active specs.
thousands died when the structure tumbled
this viscous material.
A spindle is made to spin.
1 Sedative. Made to sleep. A kind of a chloroform.
2 inserted equipment
3 load
4 charge
5 find
Loop-end loop.
On the ground, hiding.
Admiring an equipment.
A challenge for Engineers and scientists alike.
Infra – ultra.
The amazing currents of the 90ies.
Shape mimics.
Fitness.
Reactants-receptors.
Chaos disorganize > antichaos crystallize
    Chaotic   ordered
Bird on a wire:
a light entered through
    Eyes, nasal, infra.
b luminescence
c guilty
From a speck to a speck.
Attractor out.
New device.

Usability.
With the coil;
as a color detector, + net.
No ordinary thorn.
A=sentence spoken.
I A
II A-a + B
III II-b +/-
Thought outright.
Image awoken.
Word and image bound
-feelings.

Des 13th 05

It is not light, though brought by light.
From a country to a country.
Leftovers. Garbage in fact.
Inactivated in you.
In the optic hall. Search for activists groups, enemies, rebels.
Solar plant. Uplifted. Higher than before.
The altitudes.
Thailand. Netherlands, Russia: a part of the structure corrupt.
Hiding bad deeds, looking the other way, uplink signalling:
at a close range: child abuse, drug abuse, trafficking,
all kinds of pervasive sexual activities, porn industry,
women sold into slavery: addiction, lust;
just a corrupt structure. Someone slept on his guard.
-We tried to change thought
that accident in America, when thousands died . . . someone broke off.
We were running it through you.

Since Afghanistan 300-400 thousand souls dead.
Young and old, rebels, nationalists, patriots.
Arabs and Muslims enslaved in the physical.
Part drug war. Really a highly political decision.
An attempt for a Pax Magna.
Colonization of a continent. Iran partly isolated. Their tropostructure . . .
The altitudes are ours. Indonesia, Thailand. Hardly an interest now. Africa. The same.
The young of the world. The middle east, Africa. What to do?
East slaughterhouse.
A mob to ruins in Russia. The technique developed in you:
to verify users and abusers of all kind.
Both the Russians and the Chinese have profited. The diplomacy is up.
The home front. France. Problems with North Africa.
Someone sees through his fingers. Pyro-month. In England. Catastrophic fear.
The first suicide attack.
The punishment for certain crimes might be too severe.
The afterlife is just not worth it. Thus the physical led by criminal thought.
Germany. On which foot will it stand?
As a passive bystander: margin-nations must be incorporated.
The wall of Babel. Israelites. An erroneous vision? MMCrusade.
Hispanics made to open their eyes.
Problems in S-America. Youth. Population. Corruption.
The Gibraltar-route
N-Korea. Hardly a troposphere. The danger at hand though.
All too easy to build weapons of mass destruction.
(both phases).

Tartaria. Breeding rebels?
Europe. The internet, the cell phone. Secret meeting places.
Hideouts, unreadable signals. Sabotage.
To bring world leaders to one place at one time. A real daring.
India 70% in the economic muscle
Pakistan 60%?
Tartaria 10% ?
Africa 50%? (later)
Iceland 2001 7%
2004 100%
Middle East 2001 25%?
2005 65%,
We bent the steel on innocent. It wanted to go a route we didn´t find.
The Chinese executed innocent because of that.
Imprisoned brought to you. Cleansed of something they were innocent of.
And the light blazes through.
To get to know a country´s character by movement of light.
Roots of serfdom.
A discrete work.
Convents: substitute Christ where their husband should be.
How we touch the physical. May be a cause of great many diseases, hemorroid, tumor, skin diseases, cancer . . .
Good when it goes to heart for example.
Bad when it hits the eyes.
'How deep is my contrition
how deep is my loneliness'.
Here then our next appointment:
to dwell and be within as usual.
Compressing the hide.
Acuity.

Surface science.
materials
Ceramic transactions
industry
Cryogenics.
Flexural strength.
Nanocompounds.
Microfractures.
Beta angles.
Gas brought in capsules.

2006

The aura. How we touch through the atmosphere
Sense of taste, smell, lips, tongue,
Tribocha, (the third eye).
Over 200 political units.
28 m. km laser signal.
Methan and iceflakes on Titan.
Orange-bandwith in the infra.
90 km up in the Stratosphere, (50cm).
Crystal band.
Reunion. Madagascar. A kind of a Malaria?
Avian flu. Virus. Three continents now.
Cattle years ago.
Poultry.
A puffed up farse.
A new crucifix. Nails and all.
In Africa: the body of Christ HIV positive.
(Ruv(national radio), Feb. 25).
GM? Feed the world.
Fulfilled.

The hose. (Nov 2003).
Nerve-ends cut off.
Voice.
Sight.
Longings.
Hormones; no children, no virility.
Proteins, glands
Movement.
Afterlife.
Things you were deprived of.

Mars 3. 06

Devices, instruments, tools; products: you´re the inventive site
Changing the hyper structure. Losses.
1/1000 of the world population.
Run through. Impossible to trace. Impossible to
stop when begun.
Result: loose movement in the physical.
Porn. Abusers kill the longing and the joy for the body
International circles came tumbling down
Child abuse. In Thailand. Thousands imprisoned in one night.
Iranian-Russian dialogue.
Iraq: civil outbreak.
Bandwiths. Cleaning sight, or thought if you will.
New compounds. Created in you.
Semi-constant.
Semi-active
A new kind of light.
New colors of light.
Order all over.
Knowledge building up fast.

New natural law.
Usage. Eases work.
        For a structural purpose.
Inside the model.
Coloric prayers onto the wind.

Mars 4th 06

Insert dividers.
Pinkish nodes, or buttons.
Light scatters from them.
To change light, charge, color.
Burns all over.
Dermis in back somewhat ruined.
Introduced to mesonics.
Mixture of w,y,r and b, + gas.
Mesocompounds.
Divides itself/circulates.
Function of some of them unknown as for now.
Usage. Eases work in the physical.
Structural purposes.
From light to light.
We know that it´s there: we just havan´t found it yet.

Mars 10th 06

Mining.   Iridium
                        + gas.
          Vanadium
Ventilles. To divide light.
Hundreds of small units.
Some in great quantity.
Micrones.

Turbulence.
Quantizes light, heats gas.
Some optical fibres not functioning.
Light enters through
sockets, nose, through holes in the scull,
poured in to the physical.
Repellent.
Oeuf mundane
in the center.
Distinct regions,
Pressure.
Under the infra.
(Causes movement of specs).
Luminar.
Organic compounds.
Chemical.
Chemical and luminar compounds invented.
A part of the dream.
New mining possibilities.
-Has to be brain-dead,
i.e. no connection to the physical.
Your work.
Uplifted in the troposphere, (your force).
A route. A highway in fact.
Plant, factoric, (mining).
the seven deadly.
Politics. Diplomatic unit.
Guard.
´Eyes´. Some kind of satellites.
People blissed by numbers.
Edit in fact. Addicts.
Your power unique.

Sex drive.
The force within.
Work done.
To exalt devices, platforms,
´eyes´. The altitudes.
Light and mesonics.
6 colors were known before.
New colors: new combination
> new luminar compounds.
Mesonics: a luminar compound,
part chemical, part luminous.
Light, (quarkish).
Enter red, white, yellow, blue,
Exit
Green, (yellow, blue, scent of white: more powerful).
Dark blue
Orange
Pink . . .
The spectrum.
Haven´t found use for some. Others need to be perfected.
Dialectics of light.
Walls.
Light bounces off,
scatters inside a cavity.
Binds to some gases
>other type of light.
More complex compounds.
Machines built or brought.
Mining
Pottery (began 3 years ago)
Ceramic
Light

Crystals
Machine parts.
Networks.
Insulating capacities. The void.
Depolarization.

Mars, 11th 06

To make them stop
this mortal pain.
Pain and harm.
Let the ball roll.
Your kinsmen still led by hundreds.
Ordinary people, really.
Negligence, deceitfulness, lies.
To sway their will.
From a village to a town,
everyone has to come down.
From a country to a continent,
thus your mortal torment.
You lost your fancy.
You´re our pansy.
wheels within wheels.
mechanisms hitherto unknown..
functions.
Few more physical (acidic) colors. Embryo. The bluest cloud.
Emerald white.
>some physical defaults have been proven to be
caused by touch.
We built a cage.
Modified old instruments.
Muscles contused.

By mesonics and or rotor blades.
The muscles, the organs. Minced.
After the deluge.
Growing thorn-twigs.
No tonic works on that.
From retribution to absolution
:exigencies.
Denationalization: scantily populated.
Your perspiration has changed.
Your sweat didn´t smell like that.
Physiognomy: face formed and deformed.
Hard bitten and boiled.
Tottering around.
Disperse over you.
You attract it.
A kind of a center of gravity.
To be eclectic means to insulate others.
Unadorned.

Mars 15th 06

Nose. Smell.
Tongue. Taste.
Wire inserted. Compounds.
Leper smell.
Bad taste.
Mnt Atlas. Drugs.
Two routes: from Bosph through Balkan to W. Europe.
From Spain through Pyrenees.
Riot in France
New specks of light: flashy-blue (to cause immediate change of mood)

boreal green
near black.
On the far optical.
Yellow+red=orange.
Specs of white+yellow+blue=boreal green.
Boreal green+red+blue+white=near black.
Near black+orange=black,
thickness.
Charge. Quarkish.
new rayons
mesonics
mechanisms.
Optic nerves and chiasma: meshed, filtered
crystallizing on the optical.
Reduced to the cella animis, (the seat of the soul).
Pressure.
Ray through scull.
Charged with light.
Knee. Through mouth: weeks now.
Lips. Months.
Genitals. Years.
Bracelet.
Light entering through the rectum, groin, knee, eyes, scull.
Hydrolyte.
Regional mince.
Link.
Image making.
Nozzles connected to the scull, body.
Instruments brought down upon you ever so often.
Beaming.
Lasing.
Infra-ultra lamps.

Trying new instruments, devices.
Compressed in the void, (cavity in the brain).
Isolating gases.
Filtering light.
(Industry, science use mostly cryogenic, purifying).
A,Ne,Kr,Xe, low ionization voltage.
Chambers
As ′fill gas′
Furnaces.
In oxygen.
Hydrocarbons> acetylene, butane, propane.
Insulated cylinders.
Pressure up to x kg/cm2.
Air boiled: fractional distillation.
Boil-point.
Heating by compression.
Electron, muon as bar magnets.
To imitate uterus, ovary, vagina and such. Mammary glands.
Pregnancy: the loving of it: how does it look like,
is it healthy?
The mother is touched in a distinct manner.
Strictly forbidden to touch the embryo.
Reading colors. (acidic). Shortage of protein fx. : transparent, scent of gray
How to cut the umbilical cord.
Miscarriage. Our fault. How we touch the mother. How we think within. How we mark their movement. Some curious brass colors during pregnancy.
Physical illness. All to often caused by us. Proof caloric.
How the clitoris is touched.- all too popular color.
Caloric preference.
Until it stops toddling..

the Dangers.
A rebel who knows that life-long imprisonment or death awaits him if he is caught
mass hysteria
Algeria. Feudalism. Oligarcy; addictive rulers.
Enslaved youth, women, young and old kept in isolation.
Mass control through religion, etc.
Like migrating birds across the Mediterranean.
The flock has been scattered.
40% of gas in Europe imported from Russia.
A Chinese perestroika?
The Earth.
A bar magnet.
The strength of a particle´s magnetic moment.
Surveillance system.
An equipment that accelerates from within.
Repellent in void.
Reverse reading.
Movement from deed committed.
Bandwidth categories.
The splitting of an atom in the void.
Luminacy,
e.g. where people meets.
As a wrongdoer.
Look the other way.
Uplink, (signalling).
Deliberate concealment.
Hide.
Manganese (Mn) for sound
A bomb maker found. S-Arabia. Distributing arms to allies. Blast. Mostly civilians.
Iranians have a more powerful warhead. Their weaponry

somewhat unprecedented.
Sudan. Uganda.
An equipment has been developed so that individuals.
Cannot touch in a distinct manner.
The unborn of the world.
Authoritative power. Individuals are less and less valued.
Spain. Africa. Connection. A powerful corporation.
England. Someone had to go.
Hungary. Help international.
Retrieving the old imperial throne.
Enter dark blue, orange
Exit black. Made to enter immediately
>creating instruments, tools.
Scattering the scattered.
Light in through throat into cavity.
Light out mouth.
Specs of the ultra.
Rays mixed w. gases in void Ni, Ph.
Eyes. Drawing from one point, one speck of color.
Ear as mouth.
Nose.
You have been so adjusted that we can verify
up to 30 deeds at the same time.
One of the seven. Lust.
Throat, heart.
Spectral,
image search,
color search.
Mesotech.
Intro: place found.
To hurt wrongdoers,
(to make them stop).

Surveillance.
Red+orange= denser orange
(supposed to be there).

Mars 24th 06

Re-establishing your character.
Easy to change your ´memory´, as it is created by us.
In others: most effective through dreams.
-
Photonic compound + spectral color
Vanadic -
Manganese -
Force carrier.
Fervor. Infinite regret.

Mars 27th 06

Some load has been taken away from you
due to the mesonic find. New ´glasses´ for the intellects.
Everything has been rearranged within you.
The machines and tools in, on, and around you.
Since Visco-day
  Some get hurt without any reason
  some cannot touch as before.
For our defence system.
Orders in great quantity. Some units needed as they ease work tremendously.
New armory. Highly explosive. All too easy to activate. In any point-space.
(Officials have had to be here around the clock. Weaponry).
A light-bulb in yr brain. Changes light. The new light comes right back at you.

Mesonics made. All quite automatic.
It is not about light now.
Clouds of toxic compounds.
Deactivated in you.
Your losses.
We practically died from you. You were the loftiest alive.
Natural bioorganic receptors.
Something you were supposed to receive. We just have to give you at hand.
You´re not intact.
Tackling your skin, muscles, organs.
Droplets in tracts.
Light creation.

```
  A            B            F
B   D        C A E        B
  C            D        C   A   E
   F                        D
```

It has depth.
Elements inserted,
compounds form
Attach to another element
when extracted.
Inorganic chain.
Light.
A kind of a higway. Or a tunnel.
Stove.
Liquid gas.
Quite a few new compounds found. Not periodic.
Bipolar system.
A kind of a scattering,
extracted at one time or another.

Staves: mesonic objects.
Genetic structure.
mRNA codons
radiative effect, environment >evolution.
The environment ever changing.
Structural classification.
The rare nobles.
Another reason for fright.
Some just don´t reproduce.
Inactivating compounds.
Toxic waste, (in containers), blasted.
Prolonged life we care to think.
We have damaged the sky.
The atmosphere is just dying.
We have to rethink values.
Rhetorics control.
Images synchronized.
To cause vibrations in the heart.
Something to overcome.
Thorns and roses.
To level with you.
Straight from the top.
Someone is waving a new technique at us.
Scary poppers
 (embassy line)
Boiling point +600
Cubic diameter.
We lost an insider.
New gadgets.
Conductors.
Communications system.
Another compound.

Rules.
Discussion.
Debate.
A public sellout,
(to sympathize with you).
Vandalizing.
Myriads, I reckon
counterpart.
Thought to be lonely as hell.
Venerable state.
Summit.
Med. Rep.
Your belly.
Face from the inside.
Through back-heart.
Breast.
Hands off.
Throat.
Skeletal key.
No matter how much you build in the troposphere.
You have ruined the atmosphere.
In the void.
At an elevated temperature.
Cooling system.
Enter compounds.
Absorbs light.
Lips. Mesh connected to nerveends.
From the back: lungs.
Diseases, for quitters, those who hurt.
You don´t produce
photomagnetic cells.
in the region needed to be intact.

Literally invicible in the spectrum.
No aura,
no tangibility.
You don´t walk with the atmosphere.
Lost sight when we did your pituitary.
Some sense too.
Something fell apart in the tract.
Heart made to stop. Didn´t mix, they say.
Nearly choked.
Experimental harbor.
The agenda.
Healer.
        Symmetric value
Brain   heart
Left/right
Throat   rectum
Mouth   anus/vagina
Nose   penis
Ears   mouth
Every byway: mechanism
                Mesh
                Strings
Mesonics
Blue   1 micro
b.green   0,7micro
black   0,3micro
pipes, tunnels, tubes
ceramic, crystals, cubes.
Enter temple. Light.
Silvery specs
in front of face.
If this be the end of day

then man will perish as clay.

Mars 13th 06

Solar eclipse
29th equator
Israel 4?+
Gaza: gunfight 3+
Meeting of the seven about ncl prgrms
Out of Bahrain. A boat. 57?+
SW Iran. Earthquake.
Eclipses obviously a bad clock.
Deterric.
A prophetic mouth
or a generic curse?
Youth left as a joke.
No polar,(feeling), opposites.
To exercice feelings fx.
Pupils were, as was said,
drilled out before and emptied.
Liquid emptied, stored.
Gas in tracts.
Suit.
Droplets of fluid made to enter
through the eyes.
To see wrongdoers.
Other uses.
You lose inner sight.
Facial muscles out.
The region in the brain,
the top layer of the brain.
People has to come down.

Your eyes. Thousands at a time.
Light tunneling through your body.
The ring again. Modified.
Bits about the physical.
The old put a speck of light in the heart.
Cradle gifts. The see increases with age.
Another bandwith.
Phong: (they choose someone they don´t like,
Give him their worst and kill him).
The men, entelects, spirits, ghosts, national Gods.
Their baptized name used for a collective.
Your mother attacked.
Lost some precious light.

April 1st 06

-(It´s a crazy world).
High speed light.
A new thermal law.
Force nuclear.
Smaller equipments.
New possibilities.
This new instrument.
Years in development.
First time in your head last night.
Too big to mention.
Knowledge skimmed off.
Weaving two rayons together.
To feel the burn.
Toxic gas.
The body somewhat unexistent.
Just keeps on shimmering.

Strings attached.
People believes they are someone else within.
Hamas. Jemen. Those were allies.
Blast over town.
Yong and old, patriots alike.
Attack in France. Someone died.
Weaponry found. N-Afrika too.
Imprisonment by numbers.
Rebels. East Asia. Muslims. (Arabs, Perse, Indian . . .).
Mammoth anew.
1/4th of the population the world over.
Smokers. To make them stop.
Pituitary: dimensions, senses.
Balls,
light disks, (can throw elements).
Very photonic.
Highly interactive.
Nuclear force.
Splitting millions of atoms:
change of isotope.
Active haze.
Gems.
Much smaller equipments.
Better.
New possibilities.
-
Gases, (a,b,c . . .)
a acf out
b bda out
c gfea out, etc.
d
-

Radiance number.

April 8th 06

Chill all over.
Heart off beat.
Closer still.
Violence upon the physical as usual.
Throat. Fannies.
Hard cases.
Someone within you. Crooked.
20.000 kurds over the lonely now.
Ethiopia.
Lebanon. No patience. Obedience or face the consequences.
Drugs from Roman America through Trin&tob.
Transportation technique.
Multiplied efficiency.
Visco-trial goes on and on.
                    Treason
                    War-crimes
                    Corrupt board
                    concealment
                    Vendetta
                    Lies
                    corruption
Entelects all over the globe made anew.
Their sight improved. Their potentiality.
<The world Subjected>.
Qubits. (binary).
Quitrits (three).
Eliminating Dinosaurs of thought.
A ship.

Orbiting stations.
Circuit.
Conductors
Magnets
Chips
Transistors
Probes. With sensors. Collecting data.
Receptors
Channels
Generators
Diodes
Pumps
Turbines.
Nanotubes, (10 in the power of -9).
A Tesla-like revolution.
Bigger mesocompounds. 3micro.
Tens of thousands bound in the physical.
Image – word.
As a system.
Farao´s goat.
Diagnose yourself.
Exclusionary thinking.
Top speech
As deadly as ever.
Matter waves.
Micro lenses.
  Fish-eye.
  Bug´s-eye.
Sideband.
The genetic code table.
Acc to Ucl.ac.uk:
Start codon AUG

Stop codon (supposedly known) UGA,UAA,UAG
Codons>amino-acids>peptides>proteins>cell (orientation xyz)
Amino acids. Groups.
Non-polar hydrophobic (Ala, Val, Lev,)
Polar hydrophilic (non charged) (Gly, thr,)
Acidic negative charge (Asp, Glu,)
Basic positive charge (lys, Arg, His,)

CH3
Leu: CH – CH2-
CH3

OH
Thr: CH-
CH3

Lys: NH3+-CH2-CH2-CH2-CH2
Plasma (solar) particles 300-1000km/sec.
Magnetosphere: the wind, (facing the sun), compresses the magnetosphere.
Field lines compressed.
Gauss. Faraday.
Electromagnetic charge.
Current-charge-current-charge . . .

June 6th 06

Still popping up.
Dismay.
The seven. To make them stop what they are doing.
Capsules: people wrapped up.
Unit capsules: a thousand.
Image control.
This new technique.

The altitudes.
Directory. Collective.
Congo.
Indonesia, Bangladesh . . .
Quantum walls.
Globalizing administrative law.
Abstract from T*he Journal of Applied Physics*, Mai 15th 2006:
Driven by the magnetic field.
Co-ordinated systems.
Decisions-making.
Mechanic prototypes.
A motional vector: type, orientation, direction.
output
Find, save, change synthesis
Input
Axiomatic, parallel, perpendicular, other.
Nano-sized prediction pr. particle.
Determination of the signal.
Force balance.
Reverse quantum:
repel. Attract.
charge distribution.
Charge. (Ferro)magnetic.
Thermal and diffusion .
Properties of Gases.
Arc production.
Scattering methods.
Reaction mechanisms.
Building motives:
bridge
barrier
wall

Stepwise certainty.
A geometry of deactivation.
Elimination pathway.
Adenine
Purine moiety.
Planar
Steric repulsion.
chemicals
Stereo dynamics
controlled
Rotational freedom.
distribution
Charge depletion
accumulation
magnet coils. Feynman diagram.
Click chemistry: modular strategy.
F=force Newons
E=electric field (volt pr/m)
B=magnetic field
q=electric charge in coulombs
v=velocity in m.pr/sec
Electric field qE
Magnetic field qv*B>
Lorentz force, F=q(E+v*B)
A host structure.
Bubble
Spark chamber.
Wire
W-chamber.
Parallell wires, as a grid or a mesh>voltage>
traces of ions and electrons>path of a particle .
The chamber. The box. The physical

Rotators.
0,5micro.
200 rounds pr.sec.
Truth concept.
Barrel for a stone.

June 7th 06

Walking on hind legs.
Blast. Arabia. Iraq.
Hundreds dead.
Trading with Iran. Sharing technique, (up to a point).
One point locus known.
Acoustical physics, Vol 52, nr3
A waveform of a signal.
Impulse signal.
Sonar signal.
    delay line
    effect
Acoustic pressure
    source>vibrations
Thermal or electric sonoluminscence mechanisms.
Harmonic radiation.

June 8th 06

Bad decision decades ago.
Irreversible.
New device
-it just spins.
Earth science: polar (elliptical) orbit.
Orbital points.

Aura data, (to predict ozone changes).
Your aureole, (suit).
Deactivated at will.
Evanescence.
A service science.
Probability – approximations – simulation: model theory.
Design methods.
Risk: novel entities/methods
  Mature
  Amount
  Impact
Unwanted trace impurity.
Contaminant.
Promising or realized targets.
Emerging paradigm.
Foresight, predetermination.
Small dimensions.
Nanochamber.
Ethics and incoherence.
Great many women have turned suicidal because of what we have done to them through you. Some of them blame you, others not.
Jerusalem.
Over Kabeul.
Iraq. Al-Sharkawi
followers brought to you
(hit list).
Robots
  links
  arms
  tentacles
Biomechatronics.
man and machine: interaction.

Sensory cybernetics: governance.
Adaptive, artificial.
Control, automata.
Classification system.
A corrective.
  Feedbacks.
  Loops.
  Reverberation.
Behaviour-based.
Sensory, stimulant, simulation.
It takes just an instance to scan few sq. miles.
To paralyze individuals in radius.
It detects light. Individuals.
A structure.
Iran
  Blast. (Your photonic).
No emptiness in the atmosphere
-after a blast the atmosphere
sort of collapses
 >to fill up a vacuum.
Iraq.
Hostage problems.
20+ at the same place
their whereabouts known.
We gave sth to the Hispanics as well as the French.
Their Muslim population somewhat cut off from the main.
Sabo in Dk a month? Ago.
 M. East.
A religious leader executed, (15?days).
Has to be more tolerant.
In their airspace
over their sacred ground.

In you: a small equipment that has to be triggered
Peeling the skin off. Crushing and crumbling
various regions in the brain.
The spine. Symmetric.
Skin elbow out mouth. Miners.
Genitals. New ´talent´.
The directory doesn´t work properly.
New compound.
Oxygenates when out of the box. No stronger material known.
P
Gas structure He, N, Br, x, x <=black light
Mg
Used in plates, as spalt.
A shell for vessels.
Shield.
Where nothing bad can harvest.
highly active. Attracts muons.
England. Feb?
Thousands of women handicapped.
They just can´t do the most natural things. Hurt beyond.
Images misinterpreted. A lot of people was punished for no reason.
February: A structure within you fell apart.
People was attached. Few hundred women from Switzerland died.
Some lost their ability to smell.
Some losses in China, Poland, Russia.
(They touch somewhat differently. You were made for the world).
Great many have lost their sense of real in
the Muslim-Arab world.
The dead and the world. The real and the actual.
Such quantities go through you.

1/1000 just dies.
Can´t let go or sth.
Coincidences all over the world.
Accidents, suicides, anomalies.
Norway. 20-30 women died in you.
My fellow countrymen experimented upon.
The population somewhat.
Handicapped. (winter 2002-3)
Had to learn how to move again.
Sideways.
One died in the process.
Few lost the ability to move, senses even.
The see for some.
Place in the heart . . .
To split a soul . . .
RNA trace
Split, dip the trace in some liquid.
Radioactive. (Were not before).
Bird on a wire. A thing of the past really.
New technique in every entelect now.

June 12th 06

New Elans.
Heart opened. Essence from us touches your elan.
Liquid inserted. We need less elan now to bind with you.
New scientist, 90
"Navigations system.
Networks of transmitters.
Milling processes.
Model scaling.
Building blocks chosen."

June 15th 06

Vanadium>gas>liquid.
Cerebellum was the last to go.
The intermediary betw. Memory, feelings, sight, speech.
The real turned around the clock.
Before.
Now.
Valve. Genetics.
Tongue, mouth.
Inhaling toxic.
Glands in the brain.
Hot, cold.
So much for withness.
We still haven´t found use for 10-20 mesounits.
Neo-victorians: highly puritan nature.
Our children have lost their appetite for something.
Controlled
without touching
without being present
without lasing.
By the very thought. Images, loops et cetera.
Mogadishu. Fundamentalists, Rebels, patriots alike.
As was known:
the empty seat.
The invisible diplomat:
the Fatah (Arafat) and Talibans.
The hot spot.
Fluid. Metals. Gases. Installations.
Abdomen. Under the skin. Palette.
The Russian orthodox.
The Entelects of the old Soviet.

Moscow: wrong decisions-making: their loss.
Argentina: may have prevented a Coup.
Brazil: the army somewhat astray: wrong decisions.
Cuba: somewhat subjected
Indonesia: Timor. They had arms. Had to use them
Pitfalls: Ethnic, religious, political minorities, aborigines, tribes, nations without a fatherland.
Tartaria, parts of old Soviet
The Chinese government had to correct something.
Tribal war in Central Africa.
Places all over. But the world is somewhat subjected.
Iran,
North Africa,
Polynesia,
Countries south of Sahara,
some out of the route islands,
Parts of Mesoamerica.
Tens of thousands working: one goal. That every individual on this planet be within the economic muscle.
Every country has one. Established.
To do a more sophisticated search.
To make them stop.
Hurt them in like manner.
Some always die in the process. Cannot let go. Various reasons,
s.a doing this for a friend, hiding another deed, etc
Over 80% of the population on this planet in the economic muscle.
Directories.
India: few thousand women died.
Indonesia: We didn´t account for their structure: 20.000 dead.
Men and women, young and old alike.
Thailand. (Jan?) Few thousand.

-Not counting crooks and criminals.
Somalia, Sudan, Pakistan: Revolt in the making. Traced to Iran. Individuals from wherever. Islamic revolutionaries tried in outmost discretion.
The women of the world. All of them lost some talent. The Recreation of the woman physical was part wrong.
What is considered natural. They just hurt themselves
if they try to do it.
To make love, to urinate, mensural. Sth like that.
Like losing a movement in the physical. Feeling too of course.
What happens here is so irreversible.
New invention. Controlling technique.
Under the lamp-light
Morality turns absolute.
A caloric compound misinterpreted.
Polar magnetism. The energy we take from the earth.
Although it be on a small scale: this energy source used for
turbines
to ignite light emission
to trigger mechanism
for attractors, repellents
satellites
spacecraft
highways
Because of this magnetism, the equator suffers. Centrifugal.
This magnetic inventiveness.
A pessimistic approach (the pessimist then a realist)
The earth spin might increase drastically.
Tidal effects.
The moon might be drawn closer.
The earth´s orbit closer to the sun.
The weather: drastic changes.

It might rain ions, (well it does over the north pole).
Spheres might tumble .
The atmosphere might collapse. It is so vivid. Small changes in the ecosystem can have a tremendous effect on the nature as a whole.
The oceans: precipitation might alter.
Clouds. Rain. Storm.
The crust. Earthquakes of unknown magnitude.
Some great force has been tempered with.
Approximations become guesswork.
3d force negative. The fundamental system somewhat different.
Attractors, magnets for each type of light.
Polar refrigeration.
They now speak of the second coming.
You´re the fiery storm.
The awakening hour.
No slow decay.
Direct physical impact.
Blood plasma.
Mitochondria.
In short every organ.
Two symmetric tubes. Ancle.
Economy.
Self-regulatory mechanisms.
Prize.
Time.
Lim, (as tolerance: command).
Lim as saturation.
A cultural fusion unavoidable.
The mechanics a stronghold.
How this instrument within you reads light.
That´s the lock, really.

It just makes them stop.
The denser the light, the harder to act.
Epiphysical:
We just can´t move in a distinct manner.
Tamed by light.
We have listed bad moves.
A deed > an image > retribution.
Always someone within embracing darkness.
Mansjuria, N-Korea, White-Russia Oppressed
Moscow. Israel. We´re using persuasive power
Borders: China expanding to the north-east.
Ethiopia-Somalia.
Iran to the north.
Ingutsia.
Scanning the wilderness.
Falangists: bad connection.
Honesty test.
Integral of submittance.
Calculus of loneliness.
Individuals from wherever asking for legal aid.
Fundamental approach.
Strong force,
interactive,
weak force.
: Yields energy. Binds easily.
A new law.
Women opposing.
Spain: implemented gradually.
Poland: A moral revolution.
Albion´s discretion.

If this be an omega key,
then I am the abc.

June 29th 06

Until visco, you were in the now of it all.
The thick and thin of things.
Photons in phase=coherent light.
Piezo-displacement.
(Crystals. Ceramic).
Strain charge.
Stress charge.
Leak. Mossad. From the White house.
Different aspects.
Social review.
You blew the fuse.
Addicts to extinction.
A layer.

July 4th 06

Before. Cayman Islands. A tragedy.
Miscarriage. Our fault.
The directory just doesn´t work properly.
People lose more than they came for.
Too many casualties.
The Israeli went furious. Chemical/toxic from Palestine
´Banksy´ on Palestine: the biggest open-air prison.
Sexual behaviour. Reproduction.
Death pattern. As history, repeating.
A most fragile ecosystem.
Construction. Filament. Plate.
Containers: tolerate more pressure.
Light/sound/sense detectors.
Eye-plug to pupil.

Pin-point eyes.
Needle eyes.
-You have a tremendous healing power
and conversely a tremendous destructive power.
Someone died from within.–to die from pain infused upon him.
Bacteria, viruses: mutations.
The Sea: protozoa. New life-form.
Center point: universal cosmogenesis.
Smuggling instruments/weapons under ground/borders.
Kasakstan.
Burma. Structure down. Blasts.
Gliding on a platform.
Faster than a blizzard.
Calculating as ever before.
-the figures are foreseen
the numbers known.
Organs from the inside out
the world all over.
Hundreds of deeds corrected.
Habit forming.
Climate changes.

Aug 12th 06

Teeth. Function as earth.
Cut to the marrow.
Instruments.
Plugs and cables.
Through ears, out and into the back,
down and out genitals, thigh: the world all over.
Before: virtual Jurassic
-violence imposed upon

-pituitary (balance,), eyes; perpetr. Found.
Asia network.
Perpetrators, rebels, from wherever.
Children, youth.
-To subordinate.
To synchronize images.
Rewrite history.
Control.
The old of your country in the workshop.
Encapsulated, (didn´t work so well).
Amphitheatre: epiphysical.
To eliminate a deed, a thought.
As if collapsing from within.
Another child born with defects.
Canopies.

Aug 18th 06

Iran.
Building weaponry on ground.
Their control techniques blasted down:
thousands dead.
A geometer.
Light crystallizes whith the right mixture.
Geometrical forms.
Ultrviolet used differently.
Radiance too. The spectrum changes.
Venture, scope,
threshold. Pain.
Vegetative.
The body-parts man of last year. Cut his victims in peaces.
In Cpnhgn and another corpse in Sweden.

If the law cannot, we will.
The bridge between the hemispheres
is simply burnt away: man turns vegetative.
Animal experiment.
The chromosome that has to do with size fx, is just fed.
Waveguides.
Make it or break it.
Lens system.
-receptors-transmitters.
Thermography.
Many of the children born this month:
somewhat lame. Since embryo.
Med.Rep. Palms of the hands, plantar
into the bones, pith and marrow.
´Writing´on a plate.
Geometric values.
Thermal structure.
Intense warfare.
Global unit.
To be threshed upon
-or lose at hand.
A genetic key.
Image surveillance. Somewhat epiphysical now.
Your aura will be different. Won´t feel it as much..
Kepler´s cosmogony on a microscopic scale
Geometric point
How we make others do bad to people.

Aug 19th 06

New inventions every month in here.
Everything is somewhat smaller and more efficient.

Lose potency by the very act .
Thorough investigation.
The ′see′ is different now.

	length, breadth, depth
	mass
Accumulative theory	time
	permeability
	permittivity

Relative size.
Light of movement diminished. Partly epiphysical.
Iran: point locus blasted.
Political discrimination.
Extr. ′Electronics′, E.C.Young, (Penguin bks 1979):
"Electrical phenomena
Charge.
Depletion. Amount reduced: space charge density.
Relay.
Armature relay.
Reed relay.
Diaphragm relay,
-differential relays: two coils or more
    Focusing coils
    Deflection coils
    Alignment coils
Gain or attenuation.
Additive or subtractive currents.
P. 337: values of components.
    Resistances
    Capacitances
    Inductances
        :network constants
Type

Method
Behaviour/expectations"
Animal levitation. They give something to the atmosphere.
The soul has changed. Newcomers have a new kind of innocence.
So many things we think/thought about won´t even cross their mind.
Children born these last days: still born,
handicapped,
w/cleft palate. /therefore, (other reasons too), your teeth.
Viz: three figures.
Mostly from Somalia they said.
Were run through erroneously.
Med. Rep.
Short of breath.
Light out mouth as broken glass.
Smell.
Sinai. On earth. Arms from a vessel on land.
Prblms. Libya, Iran.
-(I will eliminate a movement from the world).
Angular sight.
Movement in between ourselves, in the atmosphere.
Some kind of a viscous structure shot down in Iran.
30.000 dead, they said, until they surrendered.
Opposition all over, though.
Describe: black hole. With the densest center.
-sucks you dry. Quarkish.

Aug 22nd 06

When the hormones start singing
Youth problem of course.
We can do something to the soul. Very adverbial.

The archive turned in a way epiphysical.
You gave a part of your soul to the directory.

BBC Aug 23 06

"Narrow –frequency bands.
The rainbow effect.
Chromatic exposion effect."
Sniper in Iraq.
Have not found him yet.
Efficiency multiplied, (once again).
Instead of running each and every one through you,
we run light once:
charging controls.
Mole. (Russia) in entelect.
Damascus. Opposition oppressed. (Persuasion).
Millions sort of bound through you
not bound within as before, (unless at will).
Losses. There is cost. Expenditure.
Cultural cleansing.
Popularity is not a measurement in this respect..
Light in reciprocal, (around you), by hundreds now.
Not one by one as before.
Not to touch from without. Only from within.
Image
Movement lock.
Object
People grouped in clusters, capsules.
Individuals immediately observed. Deed illumined.
Transition field.
Christian design.
Band of angels gone from you.

Eyes, (pituitary, obituary and hypothalamus).
How we touch the aura
establish truth.

Aug 29th 06 am

A container exploded over England.
Highly toxic.
More than 200 victims.
Polymer filter?
-Strangely happy before.
Through the tunnel.
-(We just saw it coming).
Caused chain-reaction in the atmosphere.
Sour rain almost immediately afterwards.
Someone bluffed himself through security.
Had made himself a gun before
showed it to friends.
They did not call him in.
Knew what he wanted to do.
They watched him. The lase.
Still reporting missing. Nearing 400
2 years ago similar incident in the States. Recorded as catastrophic.
Hotwire to make yr system 'believe'.
Your bones virtually crushed, (in the brain of course).
Individuals pored in. Loose charge somewhere in body,
sucked out.
Every so often without hands, feet, head, body even.
So strange to feel for you.
Your bonal structure, (neurons too), composed at will.
Movement of light not dependent on joints.
Glands, inner organs crushed.

The skin peeled off.
Acceleration. 10x more effective.
Manufacturing; (finding, punishing).
Thinner strings.
Different charge: supercharge.
Homosexuality now not only considered abnormal.
Punished alike.
The band: Hardly individuals. Pure thought, we like to think.
We leave our personality behind. Our elan.
Cleaning thought. The band running through.
How they influence us.
They somehow activate our thoughts. Sth base.
Bristol. Too fundamental. Islamist society erring.
Teheran. (photonic).
Ethiopia..
S-America. Sth has to be done
Entelects giving parts of themselves to the controls.
One consciousness no more a fancy.
You establish the proof, (the glands in brain; and eyes).

Aug 30th 06

New transportation technique.
With tremendous force through your skull.
As if a lattice of light . . .
The skin as a cocoon.
The cortex picked.
Light made to enter with force.
Body charged.
Movement of light:
plates
Skeleton, nerves: strings of light.

Skin: lattice.
Left thigh connected to leg, organs
Fake penis. 1/3d of the pop mask. Verified behaviour.
Your sight just gives them away.

Sept 1st 06

Raid. Transp. Drugs. From Iran
to Syria. There from to Lbn, Pal.
From the north to Russia, Nepal, Turkey, Pakistan.
Across the gulf to UAE, exp from Marokko
there from to Netherlands . . .
Thousands upon thousands addicts.
New communications system.
New bandwith.
Your spine recreated.
The twin towers done anew.
Because of improved sight.
: The subculture just knew.
Database. Efficience.
Cult around July 05 suiciders beginning to form.
Quite a few had to go.
Skin diseases.
Epilepsy: the heart.
The school in Tchetcnia.
Your sight.
New links.
S-America. Rebel army.
Thailand. (January) adversaries traced to China, Burma.
Japan. Network links:
through Indonesia, from there to Europe, N-America,
through Tartaria to the Middle east.

New uniform. Field-force.
Bandwiths: same Hz, different ultrahigh frequency.
Highly active light.
It just fluctuates like candle flame in the wind.
Iranians are manipulating/impregnating Iraqis.
Now we know how. Measurements taken.
Something in your national character.
We will break it then.
We can please a dolphin, a grasshopper, even, but we can´t please you.

Sept 06th 06

Tiniest change in charge.
An ever expanding catalogue of images.
Iceland. Attacked once more.
We feared earthquake.
Discretion can be a magnet for cruelty.
A woman died while giving birth.
The child also. Our fault.

Sept 12th 06

Syria. Few thousand killed.
Entelects DK, Engl, Fr. Drug abuse: some individuals within hiding. The last of the decadents: Backlash in Europe.
Argentina. Revolt
Amphitheatre: individuals grouped together, somewhat in a passive state, outside of.
Here: Trolls. Not normal thought/behaviour, but common.
Changing national character. Sacrifices.
Nerves: amoebic.
Circumcision. A greater come for the individual soul.

A bad thing to do.
Quite a few bands through eyes.
DK: traces throughout Scandinavia. Thousands imprisoned.
England, Canada, France. The same.
Not on a large scale now, though.
Traffickers, users, sexuals . . .
The man Deed.
Hundreds of deeds listed, not yet committed.
Possible movements created – for use outside of.
The man Shield.
The skin, nerves.
Royal houses,
institutions,
hospitals,
corporations.
The lot.
Someone magnified in U.S. crooked.
Image: 3 billions made to see simultaneously.
Argentina. Collecting data. S-America.
The same everywhere in fact. The works.
Entelects all over. New sight. Might take the load off you a bit.
3 years ago:

one lens in eye: one image

Now:

4x smaller, (due to tech, freq,), 2x (mirrors).

Argentina. A million bound in you.
World council
The power is on.

Sept 15th 06

You: energy for arms and eyes.

Agent peninsula out.
Lost a few good men.
Shooting Archies down..
Old Soviet tech.
Machiavelli marriage: to have and hold.
World council.
Entelects, magnified, and individuals make up the board.
Division by continents and by countries accordingly.
Representatives from organisations s.a. UN, EU, WTO, AI, WLF . . .
From the inventive Site.
Charge distributor.
The light just weaves itself.
Mathematical wonder.
Indochine.
On the Viscous trials.
It moved. It wouldn´t have moved unless by voltage: someone ´touched´ it.
It bounced back and forth until it broke.
Ten months later. Proof. Your sight.

Sept 17th 06

We use you in a somewhat unpresented manner.
Someone from the board. Cheating colors.
The Russians have got sth we want. Space technique.
France. One of her Entelects practically bleeding. So corrupt they said.
Woe beyond the Agenda.
Defence mechanism.
Neuro-space:
    Alzheimer

Parkinson.
´Wide´ birth.
"To-day is a thousand years."

Sept 15th 06

Iran.
-(They never saw us).
The plane crash from Pan Am Airlines in ´88
Colombian revolutionaries. FARC
Spiegel, 51/1993, p. 121
"Mao, wie wir ihn kennen, war so etwas wie euer Jesus."
Germany. Ontic problems.
Vietnam. They are developing a new technique.
Chemical agents.
Point locus.
The world all over every day.
Aleph penitentiary..
Prisoners made to work.
Search their own deeds
by continent
country-code.
To give everyone love. At night: who does he/she like the most?
Filters. (slight illusion). Outcome. Everyone more at ease.
Wormholes:a shortcut through space
Gurus: the historical need for ´a sage´ understood.
Kohl, 1994 "Konsequent den lebenden Buddah spielen, an dem alles abperlt un abprällt."

Sept 27th 06

A tree of entelects within an entelect
Nietzsche

Hitler

Heidegger

Nietzsche is one of Hitler´s entelects. German midwifes. Hid in Heidegger.
A light has been shed on some WW2 crimes.
How the magnified are chosen. Historical impact.
There were always these color combinations we couldn´t trace.
Red, white and black combinations, if you hadn´t guessed it.
The Germans have to start anew. To work with, but not against.
Their entelects all done anew.
Verbindlichen moral. Psychologische und juristische mittel.
Zauberfrau
Superweib

Sept 28th 06

Each tribe with its own curiosities, so to speak.

Okt 01st 06

A dictionary A-Z
Atlas of sexual behaviour.
Encyclopaedia of transgression.
Each vertebra at a time.
Glands, testicles, bones.
Crushed and crumbled.
Throat.
The world is not ready.

Interdependent nations.
The individual soul is charged with a nucleon. It is inserted.
-How it hurts when it is inserted. Always feel for it.
To bring them to and fro

back and forth.
New: a tiny device into every head.
No thought unless through the directory: man loses.

Oct 6th 06
Cold war ´hits´.
Paper tigers.
Blue ants.
To sit on one´s neck
going hand in glove.
Alienate yourself.
Struggle – unity.
Arguments in the line of Sino-Communists.
The political man as abstract, artificial, allegorical moral person.
(See fx ´The new Cold war E. Chrankshaw 1963)
(Communist China: Schurman and Schell 67)
Proletariat over board. Alienation, agony, pain.
Dialectics as a purifying method.
When the political man becomes corrupt.
State-Civil Society.
Generating fury.
It´s their duty to fight the aggressor.
Entelect Stalin. Entelect Mao. Outlaws that never went quite civil.
Mao, 1960 "Such selflessness (for the people) commands respect."
Japan: trade
Turkey. Free to join.
Persuasive power
Attín
Oxín
A cube

60° melts on the propellor
>mixed with air
device
Someone to see
-illusive image.
Advanced image search.
Users and abusers
-if known, in the directory.
Verified by charge depletion.
The Gaza strip
Poisonous gas. Attack in preparation.
Prevented.
Entelect Stalin.
In England. A magnified individual had to go.
The personification cult.
"Stalin, an ever-youthful shepherd", (p. 20, Hist of the world in the 20th Neville Brown).
Geometric wonder.
Structural synthesis.
Here. Filtered the ´see´ from the heart. To lead the old. Dirige.
All too easy to pass by
: illusive acts
: the heart not strong enough
: erroneous heart.
The solution at hand.
To leave no demonizer in the back of the head.
We need sth on the moon.
The Ionosphere. We can see it better now. The winds.
Feelings.
Pain, joy, sadness etc
Not only colors
they have forms.

recollection
memory images creational drive
thought feelings
sterilized

Oct 9th 06

Med.rep.
Ear
Throat plugs palms/plantars
Knee
A clamp on forearms, muscles, bones.
Deactivate. Tube connected.
Individuals marked and charged.
Man observed man, reaped holy ban.
Flashpoints. It just oozes.
Central Europe.
Old curtain countries. Burning altar. The lot corrupt.
Orphaning children,
widowing wives
Periferi.
Space tech.

Oct 13th 06

Over a year since visco.
Mostly from Louisiana. Just days before the hurricane last year.
Last ten days. Taiwan.
Preparation. Finding, marking, charging, running through.

Oct 16th 06

New elans.

A tenth of a normal portion
30 Entelects
60 countries
As groups, (starboard). Not movement.
Mongolia.
Run them through.
Do something here at the same time.
<We can rearrange our genetic code>.
Splitting the split.

Oct 18th 06

Attach.
Elan + compounds.
Morning. Some newcomers died from you.
Evening. Few of them put back.
Eyes. Always the eyes.
Communications system.
Technology to block, disturb, scramble . . .

Oct 19th 06

A historic freeze.
A layer. Thickness; few meters.
A rare element needed.
Compounds.
Substitute.
Not found
not known.
The will and the way.
Before. New elements found. Very active. Unstable. Isotopes.
Here: your people. Your losses.
How they move, hold onto, hearing. U-h-frequency.

You are not likened, to say the least. Hated by a definite number.
Various reasons.
Addicts. Imprisoned or executed.
Homosexuals, lesbians: fought with vigilance.
Punishment for so many things that were common.
Women: have all suffered. So many have died. All too many handicapped: women that
Cannot enjoy sex-life as before,
The seven deadly
Controlling technique. We are made to feel at will. Thought control.
The opposing voice is veritably out of it.
Directories. Some have lost sense, movement, . . .
The war. Subjecting nations, enslaving their people
Mistakes. Embryo, (has been prevented, we hope).
all too many have died when we were running them through you.
The Viscocity trial: 14000 Americans. Mostly from Louisiana. 12000 immediately. Others in the weeks to come. The most painful death. Your fellow countryman, (executed in July this year), touched this viscous material.
Those who knew. Those who didn´t. The secrecy about it all.
The clouds are somewhat dirty, (compounds).
The sun filtered
The winds hantered with
The sea: we tunnel sth to the sea. To try to save time. As a garbage disposal.
Physical. Lines of force.
Fx your hands.
Inside yr head
: to create
: illusive; feels as if from the outside.
China. 10000 killed.

Persuasive power.
N-Korea isolated.
Embargo.
Directory problems. Few died

Oct 22nd 06

Cultivation.
Changing society.
Bad companies and good ones.
Changing market value.
Direct or indirect influence:
money flow
sources
New knowledge. New plans.

Oct 25th 06

The Man Ingutsean.
Eastern Asia.
Uganda. Chemical Weapons smuggled from Iraq. In the hands of warlords.
Mao. A part of him died during the attack.

Oct 28th 06

Glasses. Phenol acid.
Chemical in blood.
Center of brain:
trying to simplify.
So that the system works.
Feelings – form.
To hanter with the helix.

Memories.
Thought.
To will. Not to will.
Acidic colors. Every feeling has its distinct form,
(happens in the plasma, outside of the helix).
Just bypass the messenger.
Molecular mixture:
proteins
gluons
A real daring: you have to be experimented upon.
To feed others with feelings.
Last week. Emphasis. Stomach. Overweight. Our thought in a way.
Lungs: pneumonia. Our thoughts in a way.
Hemorroid.
Here is a story
Bolsoy gory
and his lot
Deserved what they got.
30.000 killed from Russia
Mai. Kina. 50.000?
Charlie company

Oct 29th 06

Sion.
David ben Gurion.
Ordered Beirut attack this summer.
Interference in Nepal.
Lost a branch.
War crimes. Some who were said to have died during the holocaust

really executed by them.

Oppressing neighbours.

DBG Tree. Einstein within.

Turkie English The man Cairo

Italy American

Ambassadors, magnified, individuals:

few thousand individuals.

DBG within Sharon. (was made on deathbed. (highly unusual!)).

This strange intelligence.

Leak from Mossad (iyw), i.e. DBG through Man Cairo to Russia.

Einstein´s idea and calculations decades ago. The layer.

Raining Ions.

Oct 31st 06

Japan

Warning shots

Allying W. China

Indochina plans

Mustard gas in Arsenal

A tree

Their man. In Phong, an American-Japanese

France

Spain

Chile

Brazil

Seoul man

Stalin

China

<a web of lies>

Trade, as was said.

Chemical agents. Think differently.

Just suck them up.
As before: They cannot see the enemy, (us).
The objective is closer yet. That the men will all be under one hat, so to speak.

(you hold the hat)

Splitting the split
Every prisoner
Every newcomer

: we just see what they see.

The bombings in Lndn, July 94.
Not IRA. Muslim extremists.
Man Ingutsean brought U and Pl to N-Korea.
Japan. Attacked us. Rather die than be a part of this.
No casualty.
Taken by force.
Lies about WW2
Organized economy
Markets. Causing countries to slump.
Before. Sects
Women Africa.
The ′circumcision′ takes pleasure away from the physical. Some cuts in the brain and the orgasm is more powerful for the soul.
Forbidden fruit. Those who mutilate the body in like manner lose their place in space. They get death sentence.
Men circumcision. The same, really. Less to the physical, more for the individual soul.
All too common. A mammoth thing to do. A society has to change from within.
Polygamy: all too common. In all too many countries. Certainly not advocated.
Shia, Sunni
Approach the physical differently. Architecture: slight difference.

Old rivals, really.
The cosmic route changed again. 30mpr
A week ago? Cosmic clock
Chemical, atmospheric sort of a collapse.
2x a year?
To die in
To die within
To die from
The Coptic secret.
Inter testamental times
The fiery monster of the Balkan high

Nov 05th 06

Med rep. Blood w. sperm. A small vein in penis gave in.
Violence over and beyond
Mao out. ′Reading’ Mao. All over the communist world. Diplomats lost.
Italy. Eritrea.
Through Egypt
Traced to S-America.
Closed the Panama Canal (from above of course)
A container. Blast over China.
We go away from China now.
Sovereignty
Defence circle
Image of a cop, or a soldier,
Appears close by suspects.
They don′t know where to attack if they wanted to.
We can see from there. Shoot from there,
The bridge went down
Magnetic defence

Nov 7th 06

New instruments every day
A technician´s delight
The world all over.
Not by towns or villages any more
Only those who have done a deed or another
Bad company list. Run erroneously, by crooks, etc
In head. A Ring. Filled up w. glanular products, differently charged particles.
Enter: individuals. Exit marked individuals. By deeds committed.
: deed illumined by charge depletion.
Nursery: watch them closely.
0,001 nano charge/voltage
A nanoscopic sight. No small revolution.
A thousand equipments have been tried on you
Some have not worked properly, others have worked wonders.
You´re the mine of mines, too, then.
Ramifications
New Calculus
New arithmetics

Nov 9th 06

Hamas, Fatah,
Sinn Fein, even.
Not so much drugs. Patriotic, rather.
Here. The ceiling established.
Palestine. Ceiling. Israelis oppose. Persuasion.
They´re gonna have to control their people, as everyone else.
Not so much by nations.

Diplomacy must up in this part of the world.
The molecular structure of the atmosphere a 100 years ago may have been a bit different. So radioactive under and around the layer. Chemical changes. The cosmic clock. (makes us all afraid).
Precautions
:Changing the cosmic route may have delayed it
:Created a compound which is not as active. A sure delay, (if calculations don´t fail).
:A kind of a canopy. Magnetism. Electricity. Other energy sources. Infiltrating the sun.
The light from the well
Eastern Europe. Just sucking them up. Addiction. Sex crimes. The same as everywhere.
They still want their mines
Vietnam
Iran.
>persuasion. We have to agree.
Filters. Moon. As if it reflects sunlight better.
Illusion. Just light from us.
Pearly rivers
These last months' new laws of nature have been discovered.
New energy source. We hardly know the consequences.
As before: the pessimist and the optimist.
Right or wrong calculations, etc.
-Proto-compounds
With elemental moments?
New satellite (surveillance system, communications system, . . .) thrown high up on orbit.
Distinct organic compounds
Gases
Fundamentals

(a highly active element fx.)

Inorganic, organic elements
We ruined the communications system for the communists:
China, Russia, Their allies.
:-We have to agree.
New techniques>new possibilities>new solutions
The nano spectrum
Mass depletion.

Nov 9th 06

Med rep. Enter Gases +sentients.
Space suit. Aura. Glands.
How we move while we hurt
Respond. Sexual behaviour.
New compound
Didn´t bind on first row
The stock market. Highly overvalued.
Colombia
Haiti connection.
Missiles. Bought from Cuba
Here: youth. From anger to frenzy. Made to feel.
Amnesty International. Some of her groups are sub rooted. All to influential. Looked into. Who supports her. Her decisions making, etc. Individuals put themselves all too often in the line of fire. Activism negative.
:cause anger where there shouldn´t be any
:disturb a peaceful process
A good cause craves attendance
Prisoners of information
Defloration
Last months. Few thousand girls lost their virginity here. (The youngest seven).

Most of them from Spain, someone said.
Couldn´t count them in, as was said. Such great quantities. Such erroneous deeds.

Nov 12th 06

The arts, from a point arbitrarily chosen, either hides or reveals sins.
Med Rep:
R.eye. Gas enter
Hands: gloves
Knees
Left ear: enter, for months now.
Mouth, bones, genitals.
Hands. Fx in head, in body
:A cap inside cortex.
Skin. Hair. From front (heart), out the back.
Int. org.
So many deeds will be run simultaneously
A crooked American in one of the Russian men.
Cold war crimes
A mole in fact.
Old and high up.

Nov 14th 06

New elans
Man Portugal: Tree
Southern Europe
Brazil: Sao Paulo
Guatemala
Nicaragua –Panama
Macao

China
N-America
England
Pakistan: youth run through.
Iraqi man.
Amphitheatre. Bigger than any a stadium
Tropical Africa
Ethiopia
Tree
Ambassadors
Papacy
Cairo
Very authentic. The Royal house. The parliament. Parts of the army.
Need new technique.

Nov 17th 06

Lines of force. From within
Bloodied week. The deeds we found practically everywhere
DNA hantered with.
To take away talents
Sexual abnormality.
Lose potency.
You guard the belt of chastity
Feelings as forms
Hatred. Just fill the need
We run them through you,
Every so often summoned in their own courthouse
To give or take talents has a new dimension now.
More difficult to control addicts in like manner.
Not as base. More complicated

Essence: yr elan + essence from us
Essence in pituary, obituary, in heart . . .
What do we love or hate: their hide, wrongdoings, accordingly.
Rephr: what we like and don´t like –for the physical.
We now hold every entelect in heaven. Not every propaganda machine though. (entelects, whose heart still ticks).
-(Yr knowledge is kept in another format).
Bits and pieces
Food factories
The isle of green
Drug related
China again. After a short recovery
:Incest.
Teen prostitution
The same, really.
:you will feel less than other beings
Tibet. Well over 2000 monks practicing homosexuality. The population seemed indifferent. Made us think. Drug addicts on the other hand, (Nepal too). Their chosen one just corrupt. Had to go.
Abnormal sexual behaviour on the one side of the scale. Addiction on the other.
On the war in Iraq. They had old Soviet arms, but started suddenly to throw weapons up from their man. Was killed during air-raid on ground. We lost few thousand angelic soldiers. An entelect (American) died (Nov 05) The last of any serious casualties. The technique we invented in the following months secured our superiority.
The Iranians couldn´t intervene. The French would have joined forces. Also: German diplomacy. Therefore as silent bystanders.
The Iraqis make their own now. (under surveillance) Built for peace.

Morocco (early this year) One of their men died during tomie.
Another had to go. All too addictive.
Prostitutes from ever
Organized crimes
Although drug trafficking and sexual abnormality may have diminished on ground,
new ones take over immediately. To be hosted with two or three addicts, even.
As was mentioned before. Heart problems. The cradle gifts, resp.

Nov 21st 06

New tactic. Human shield
The Indochina peninsula
Vietnam. A man in the making. (again).
Detectors
    Activity
    Movement
    Sound
From the East coast of Russia, Vladivostok. Trafficking heroin to the west coast of U.S.
From South America (Venezuela) to Vladivostok. Cocaine. There from to the west-coast.
From the solitary north

Nov 24th 06

The vermiform appendix
Gastric acids
Scrotum
  Central nervous system
A ring
Magnetic

Active
All of East Asia. Joint effort
Treacherous Ghosts
  The American-Japanese
New defence system tried
Internal affaires
1/5th of N-Americans from Europe
2/3d of the English
  The American-Hispanic
  The American-Russian
  The English-Indian
Ghosts acted on an international level, nationally.
That will soon become a thing of the past
The Japanese will never forget
-you rode the Enola, then.

Nov 25th 06

Russia. They have a deadly laser
Don´t have a nuclear
Their men as ours, I guess. One goal,
fortified defence system
Throwing arms up from you
Improved network
More mobile mechanism: ignite at will, not ever glowing
They changed their route,
Can lighten up their heights, in search
If that was their best, then we have them (for now)
-(Give me two weeks to make a man: if I´m not to worry about the man´s wellbeing)
The difference between a tuna and a whale is wellbeing
A kind of a covenant. Not on earth. That would be the last

resort.
Play diplomacy with two options: surrender or attack. Goes for the covenant as well I suppose
Sacrifices
On the Americo-Japanese
: we really were closing in on him (others too)
They were buying allies
Billions of $ to countries in S-Am, (Indochina peninsula)
Neutral countries get a sceptic´s look
Verbal carpentry
Milton:
"The firmament
Radiant clouds,
Orion light
Crystalline seas
Golden hinges and compasses"
I guess we stepped on too many toes these last months
As God´s eternal store
Dizzy ghosts
A glimpse into near future: pan-slavism
Don´t get caught trespassing in some
Quadrivial and quintessential place
A part of the board
Sided with the enemy
        Short sighted, egoistic, personal
Locii: whereabouts known
Not only to find the men, but actually participate
Aerial museum
Equipments
Romans: army of slaves
To bind the soul within the body
Mediaeval punishment

18th century: to burn the soul within (fasten)
Was still practiced yesterday in some
remote places of the earth.
On slavery in the Americas: They left nature spirits behind
Made by white man´s light. His burden.
Face off (Beginning of April 06)
Indiv. Made to believe
they are in someone else.
:Recognized as someone else
Japanese WW1 man
How corrupt these ghosts had become
New trial.
Now F. Roosevelt
Here: before. The individual soul was left unpunished,
Even though there were eye-witnesses. Burden of proof
(We woke up one day and there was prison, inquisition,
death penalty).
We of you who are young have had to convey the
meaning of the elder.

Des 2nd 06

Tens of thousands Arab addicts imprisoned or dead
Preparation: been marked and charged these last months
(lied to them.)
They won´t run the show
Machine in head
Back of brain, inside scull
Cavity in between
Filled up w. liquid (fuel)
´candid camera´, =200 places at the same time
Seen from within

Wpns marked (in the inferior)
The opposition takes the markings away. Didn´t work
Nclr power plants
We´ve got some structure there
Ghosts. One by one
Deeds Done anew
Lose some sense
They get attached.
The individual who acts (sum
Total of thought, iyw) directly
linked to the board
Drugs from DK by sea to Greenland, there by flight
through customs Canada
Freud´s dream dogma
Manifest or latent
Vivid hallucinations, used by addicts as ´purifying´,
i.e. to hide bad motives.
Difficult to interpret.
Not until they´re a part of the dream works
100000 within at the same time. Addicts.
Marked before. Passive. Those who lose charge . . .
The man S-Africa. Found corruption high up at home
Old tugs
-Nazi hunter lied
-Who leaked the combinations (WW2
Communic. System)
A part of one of the man French helped subiecting France
Same with Belgium, Austria, Pole,
The Vatican under Steel axis heel.
Some of yr fluid from ocular tracts
was cavitated before. Same with nodes.
Lost sth precious, rem.

Sight burnt out (April?)
Droplets in tracts

Des 9th 06

New elans
Breaking the defence walls of Ch, Russia.
Fortifying ours
Industrial espionage
New checkpoints
Codes
Constantly changing
random
Here. This machine of theirs. Quite a few elder died.
Dialectics: Dogmatic, critical, genetic
Bias (voltage)
  Particular, characteristic
Oscillator
Christianity.
Feedback. Ghosts all too adjective
The wars of last centuries. Ghosts vs. Ghosts.
Warfare their bread and butter.
Communism
To feed the poor. Stands firm feet on Christian ideology.
Their Ghosts, accordingly, Rascolnics
Insiders, moles
Espionage and like
Resistance, struggle, strife
Conscience a fact, concomitance not
Russia: old soviet tech they never used
        Netw. System,
Hammer and sickle

Des 11th 06

Times square countdown:
Net points
Attracts light
from us, both upward and downward
Loss: physical loses some sense
Traps wrongdoers
Swan fight on the pond
Bias, broken, meek
Never sick, always weak
Before evolving, primitive
Recoils
Reloads
Obvious joke-avoidance
Ceremonial prohibition-punishment
Obsessional obedience
-Kadesh-
God of the old testament
Chose and forbade
>dismantling
Of wanting to fail
Six white roses. The killing of the prostitutes in or near Ipswich:
Made to happen. One of them had AIDs.
The man German WW2. Rommel?
A part of him in Churchill root
Extr from ´A portrait of a man standing
By Salvador de Madariaga, (Allen and Unwin Ltd ´68
P. 56 On German Nazism:
“the shepherd of that herd was a vertical person.
A true upside down. Blut und Boden”
Russia-Muslim man

Syria
Hungary
Algier
China
Muslim population in Russia. Dialog w. Iran
‘ The 20th century in Europe’ by R.S. Latourette
p. 293-5
“In the Nazi period he spoke of Jesus as ´führer´;die führervollmacht Jesu.” (in the words of Karl Heim (+1957)
Rudolf Otto (+1937) held Christ to be the Crown of Hinduism.

Des 17th 06

Iranians preparing. Prevented.
Russia. Once again one of their men
The third Man German from WW2. Hid too well. Freud spoke of him, but we
couldn´t find him.
In Tibet
Romania
Hungary
Yugoslavia
Weeks before. Mexico. Ghosts fighting. Had to interfere.
Convincing individuals
Whole of Russia just corrupt. Control fiercely.
Make their Ghosts anew
Persuasive power
The magnified all over
Part of the WW2 white list wasn´t so good. Magnified w. their men, individuals not likened.
Our Ghosts. Their Garments somewhat dirty too. All of them lost some individuals; addicts, homosexuals, murderers, even.

Our magnified. The same. The Girl´s best fx,
On earth.
The men done anew of course
New sight.
New men.
What sight!; they say:
you are made w: old tech: you got the brain, though.
(Categorically brain dead)
The world bound in you. Rectum, tessies, muscles,
tissues, bones.
Could correct sth from WW1: Ghost vs. Ghost. Images pair.
Sth went up with Discovery. On Solar orbit. Experimental purpose
Our man Indian (N-Am)

Des 20th 06

We now can extract glanular product from diseased corpses.
This new device. Not so much for our individual souls, nor to control them as such, rather: space project. To attract
at great force,
to emit..
Androgen and estrogen
Their base the same.
You have seen some equipments. But this beats them all, we think.
We chose someone to experiment upon. (Guess who).
If it works here . . .
Ocular fluid
Might be able to improve your sight a bit.
Make you slightly intact
We can insert some tiny device in a man´s head, and he acts as we please.

No intermediary
Something we did to us all
We didn´t know. How to explain? This generation
Cannot love as much. We lost some physical sense.
Horrendous, fearful
We dare not try it out
Wld be global delete.
The milky way seems so much smaller now
We are faster now
More and more epiphysical
Just adjust for the right attitude. Switch on.
A kind of a lattice. A film.
Between us and earth.
That sort of wraps it up.
A cosmogen creator?
A God of no testament
The grand design.

Des 24th 06

Somalia
Their man was made differently.
Mostly by magnified from:
  Libya. Weapons aid
  S-Arabia
  Pakistan
    There from to Indonesia.
Kill innocent
Burden of proof not present
Amputate
Steal women
Woman Arabic. Not as corrupt as the woman in the W-hemisphere.

Partly because they have been so controlled. Partly character
Homosexuals in Muslim countries. Hardly existent. They just kill them.
China. Fewer than in the W- hemi.
Here: we now can up heave children much younger. Few weeks old. Couldn´t before, and wouldn´t have: too much work.
Your Elan Vitale, if touched, somewhat prolongs life. Those who ´touch´, inherit your soul: we are you.
Nodes in heart. SA node (sinoatrial node)
The physiological ´pacemaker´
AV node (atrioventricular node).
See fx The Atlas of Physiology. (Despopulos, Silbernagel)
The French educated scientists from Iraq all throughout the 80ies and the 90ies. Lies. Aid: monetaire. Advice.
WW2 Why the sudden Change inside Soviet?
USSR and USA let their men change ghosts.
Mao and Man USA. Changed diplomats
Romanov execution. The Ghost responsible known.
Russia. Others too.
Didn´t show them until now sight atmospheric.
Had proofs on few thousand, murderers, addicts and such
The Man Finland
Neighbours within, as is common.
Addicts, Homosexuals.
Didn´t follow suit. Took the bait. Told Russians.

Des 25th 06

Bangladesh
Bahai. 40 millions?. Somewhat erroneously run.
New sight (in a way). Atmo, (left is right &vv) + mirrors
Individuals around and above just seen.

Med. Rep
Deformed
Skin. Back
Your throat. Tried to repair it
Mining: Compounds. Fractions of a second in the making.
Equipm, instrum, devices: minutes, an hour, few days in the making.
Light and gases just pouring in.
Old and cold cases done anew
Ghosts, magnified, individuals

Des 28th 06

Tremendous blast in thin air. Yr device of hundred possibilities.
One hand up against one another.
3km R.
All kinds of small devices made from within. Heat. Melt. Metalloid+gases, glnlr prdct.
Prisoners from Somalia
People bound and rebound
Here: thousands marked. Sth set within.
So that they don´t want to
Hantering with the souls. Building a structure

2007

Barth, K. 'The humanity of God´ (Fontana ´61)
P. 40-41
"To make God great at the cost of man . . . that God is everything and man nothing . . . the result of the new hymn to the majesty of God a new confirmation of the hopelessness
of all human activity": God is over and against man.
46:"God´s deity is thus no prison. Rather His freedom to be in

and for Himself
But also with and for us"
44 Mediator, Revealer, Reconciler
45 Christology: the hypostatic union
The Condescension of God >exaltation of man

Jan 2nd 07

Some of the Ghosts have practically become like the Green knight. Their own filth
Their individuals profiting, pleasing each other. Monetaire. Did sth only Ghosts could do. (Well they are more powerful than any particular individual. Know all the tricks in the book, etc)
Barth´s doctrine of Togetherness.
To bring this further. Togetherness of Ghosts.
Med. Rep
New Equipm. Testicles to incr. glnlr. Product fx, oval. Up through the abdomen,
muscular bound. Throat, mouth; imitate vagina, anus
:to make them stop
Mostly womanish
:Cleanliness: (semen)
Have to hurt them.
Few months for sure. 30 deeds at the same time.
The world all over.
Made to cough.
Hands, toes, limbs, tongue: light of movement. Lines of force. From within.
Face. Dermis. Epidermis. Muscles, stomach, interior organs, genitals, anus, rectum, glands.
Did I forget some?
Light. Strings, Strays, Specks, Fuel, gas, compounds, impersonals..

USA.. Internal Affairs
Individuals within Ghosts, magnified; giving each other's inside info about companies,
organisations. Few thousand. (All of them white) Giving to friends and relatives, inviting each other to join their men and their magnified. A considerable portion of the groB national.
They were not in it for a healthy workforce, nor to fight any of the woes; poverty, famine . . .
The Old greed in and for itself. Ten of the top hundred richest.
How secretive they were. Signals here, a word there.
Here: Someone within. Gave friends and relatives the best of dreams: language programs, history lessons, news . . . ignored others. A ministry turned suddenly available: a relative got the job. A director job in a bank. The same. Things like that.
From the center to the periphery
:there is always the individual decision.
(when the individual turns out to be within a Ghost
inside a Ghost: A root).
The trance. None look back
-give them the time in it.
Your heart
How we attack the heart.

Love
Control
Drive
Bypass the ′see/feel′of the heart

The worst yet. People who wants to hide.
Heart diseases
Few new colors (acidic)
Miscarriage. The first two months.
We just see it pump. With your sight that is.
Yr. heart. Strings attached

Its place in the brain
The glands: ob, Pi,
We gave our people something. They just have to give it back
Sth we have done since the 17th century. By then cradle gifts became more elaborate.
How we hurt the animals
We help birds migrate
The young. When they don´t like something.
What form does their thought take?
Feelings we don´t like
How we reflect/respond to stimuli
Geo of possible forms
A complete list of pure geometrical forms
Derived
Manufacturing feelings.
In a way excess on stock
A box (physical)
Glanular product
Biochemical product
Compounds
Molecules
Protocompounds
Glasnost: transparency
Perestroika: reform
To see how it feels
Grades: just a different form
Birds
Animals of the higher order
Primitive or advanced
Difference: adaptability
Radiating forms
Convergent forms

Some barbicel family
A taboo before God
who calls bad things
upon them
Gulag
Abduction of Baltic families
by the Soviets, ´41, ´49
(Bartholomeus night)
500000? Mass graves
´The fontana economic history´, ´76
Hirch, Oppenheimer:
The Marshall plan, ´47-8
To keep it recycling
          Currency transferability
No Quotas or restrictions
On non-agricultural trade
Sterling Devaluation ´49
Franc devaluation ´58
>healthy balance
Keyne´s advocaton
Friedman: The invisible hand
´once elected, the right people
do the wrong things´.
It´s the end of Ghosts as we know them
The crucified allegoric
The cave empty allegoric
Biblical error
Political truth
Historic loop
Passim, polemic
A Ghost as a kind of an
Angelic soup

Hitherto not a tasty one
Their decisions-making
Throughout the centuries
Why so corrupt?
Know the trick of the trade.
Hundreds of deeds were improvable
Two years ago. Proof of burden
Cultural sickness
Lazy wellbeing
Uncleanliness
Cruelty, greed.
Vengeful, deceitful
An endless list
Heart.
Erroneous task
The world changes fast
The very physical tone
The opera magna
Your pain to bear
The men who come in your wake
will see better. Ocular tracts will
be more efficient than before
The next one will be made differently.
More specialization.
For air trafficing, fx.
Ornaments and vanity
"There is something
wrong with the dice"
Jevtushenko, ´67
Penetrators: send data back
Penicillin and a specific type of flu:
Ineffective.

Shock value
Shock tactics
Expression. Attitude. Comfort zone

Jan 7th 07

Accident: over Europe
So dangerous chemicals. Few missing

Jan 10th 07

Année Royale
Few hundred millions
Imprisoned by you
Togetherness
Our old Ghosts come together
As one.
Probably the biggest day
In the history of mankind
Turkmenistan. Administration
Some peptides don´t work in you
Have to insert them (body parts)
Not the same though
Also: enzymes
Haemoglobin: a kind of an iron-peptide
New element
Not postulated
Captures photons
A Compound. Few hundred elements,
Skilfully constructed
When active: protocompound.
Point lumens. In reciprocal.
They just give themselves away.

A forgotten tribe
Central South America.
N of Amazon
Problems in Polynesia
Synchronizing images
Decision-making
M-East
A civil outbreak (on a small scale)
We just make them run back and forth
Until we have found those we are looking for.
Somalia. A pirate broadcasting
Old Curtain countries
Riddled with old communist thought.
Slowly but surely changing
Any traveller of ours
Turned the canopy off for a while
To smoke wrongdoers out
Mongolia: the better part of her was in the hands of Mao
Russia. Breaking apart?
May be we don´t want to
Has to be run differently
One of our old
J. Wayne root. Born mid 19th c.
Existent in him and Roosevelt.
Sub rooted in England, Germany, France,
Intensive care
Steel industry will go downhill
We know of a stronger and lighter material
We just need to blow the pipe.
FN in France. Not so much extremists (any more)
Should have listened to them at a point in time

Jan 11th 07

The world bank
Not strong enough
Governments of participating nations
Must decide on its future. Incr. cash flow
Similar organisations. Similar response.
These changes have been
terrible for us all,
Young or old
Wise or ignorant
Guilty or innocent.
Always some casualties.
Something we didn´t know before,
element of surprise,
mistakes
Our space-project
Go to another solar system
-and live to tell
:Building a new space-ship
On earth: this Promethean fuel
Hitherto: needed the physical to survive
: to encapsulate proteins, enzymes and such
Our not-so-old solar plant
has come of age
Our new tech
New kind of waste
Evolving problems of course
200? Individuals within various Ghosts
In the W-Hem had to go. So corrupt.
Russia Mob
Laundry cross border.

The eastern bloc.
Whitewashing each other's clothes
Proof of burden in the actual.
So difficult to trace.
Privatisation happened too fast
Without surveillance
Ministries of finance in some e-bloc countries
Under scrutiny.
Drug profit
Illegal activities
Officials
Countries where the Roman Catholic church
has the stronghold erroneous.
The History of Christianity, by Latourette, Vol V, 295
Historical. The exodus of missionaries, Europeans from N-Africa
292: Suez crisis ´50ies.
Sequestration by the Gov of Egypt of the property of foreigners,
Incl. missions: departures
Missionaries:
Conducting hospitals, clinics
Medical
Education
Orphanage
Propagation
290: because of Muslim law: penalty for apostasy:
Death or at least social ostracism.
Centuries-long dislike: few converts
1950ies. Franciscans had 38 parishes, 42 elementary schools,
Five secondary schools in Palestine, enrolling not only
Roman Catholics, but ´dissident´, : Christians, Muslims, Jews,

40yies. Masonites were predominant in Lebanon.
Chaldeans, (nestorians) on the planes of Mesopotamia
P. 288: After the late 50ies turbulence in M-East
Secularism made inroads on Islam. From pride of cultural heritage.
Nationalism quickened through conflict
>exodus of Europeans, missionaries
:vented their fury on Christians
In countries where Islam has been dominant
Politically and numerically
For more than a thousand years:
Whether Roman Catholic,
Russian Orthodox
Protestant
Or some uniates,
Bahai or Jehovas;
  persecuted
Revisionists and their nightingales
The man Africaan holding the Dutsch population in S-Afr (Transvaal, Natal, Orange)
Very strange community
Very racist
Very few mixed marriages.
Not so strong Easy to fool, iyl.
Easy for those within w. the know-how.
Swaziland. Much more tolerant. Drugs. Inter-Afr. Links, Portug.
The man S-Afr, (made by the Commonwealth) within the man Swazi, the man Africaan, and vice versa. We were not so nice either.
Big Sister.
The Roman Catholic church.

Her interference in S-America, Papacy, Portugal, Spain; countries in Latin-Am.
One within each other. South-central Pan-American reverie. Bolivar-dream, iyw.
Masonry link. French.
Protestants. Few in the west Indies. N-American Methodists, Baptists, Anglicans.
Old Soviet´s, and now Russia´s ´Muslim protector´, Their man.
The break-up as a scheme. The economic factor was their neglected war-zone.
Russian and Greek orthodox, Big sister and socialism.
Gibraltar strait.
You: input phenol acid. Causes chemical reaction: instrument ´glued´to its place
East Berlin syndrome.
Japan, China, Koreans.
We turned pessimistic suddenly.
They figured out a defensive.
Situation extremely delicate.
How sick we had become
Socialism as of now
National resources and economy
This protestant alliance
Has put the world on a brink again.
More and more automatization
Stalin root in Castro, The man Nicaragua
Sexists.
The jet-set once again
Hurt them to make them stop
Equator
Surinam
Scaling and spacing (you)

Thanks to you, we are, each and every one of us, the same, from the pith to the marrow. That means we stopped
doing some things.

Jan 18th 07

To use WW2 analogy
A quisling. Short trial.
Russia. Sent up a lite. Spotted our flying carpet. (our platform).
Shot it down. Few casualties.
Few thousand in Russia. Iran. The Russians came to back them up.
Our equipment of a hundred possibilities.
Hundreds of thousands. Unconditional surrender.
The man Iranian blasted.
Probably their high-ups.
Russia: surrender. Cost: their man marked for us to see.
Lite down.
Center focal: you. They know where you are
They have seen you. Your country on the line
The blast was so tremendous that part of our troop across the border died. Few hundred.
Spain. Latin America. Cuba
We might lose goodwill of friends, allies, even. They speak of ethnicity
The blast, it is said, was bigger than was expected
Chain reaction
Reverse storm
They tried to run away. The shock got them
The Device. No bigger than a horse´s shoe when inactive.
Someone within Disney.
"Diplomacy". Russians agreed w/modifications.
Let´s see how that goes, then.

On the war in Vietnam. We might have prevented it if we would have given way
Somewhere else. USSR gave capital, weapons technology,

Jan 22nd 07

Digital sterile
From the world front
Not to kill innocent
They are in advance in that respect
Maoists, communists, socialists . . . and parts
of the Roman Catholics, Buddhists, Muslims:
half of the world population
One of their man. Dead. (Strong but not big)
Ancient. We have been had since the beginning of
the cold war.
The Jews shaking hands with both parties.
“A broken handle”
You: Sedated. Ins. Chemicals
Individuals marked
Another lethal. Various utility.
Looks like a door knob, when inactive.
The physical as a plasma; we, the light, our equipments . . .
penetrate the softer materials.
We have been misled for a long time.
Something we do to you that turns so physical. Violent, even.
If you move, someone hurts. Hits you back. Turns violent.
Few new platforms.
Gliding.
You ´see´ for them.
To see their move before they can synchronize their thoughts
The holy see throughout history

To rearrange truth. We tried to make every living being
forget about you. To recreate you in relation to others.
(You were sorted out, enveloped).
If the individuals buy it, the physical will; directly or
through dreams.
Hear my people shout and cry both from within and
in a distance.
It tumbles. (Our thought of you), once in a while.
Sth in the actual reprimands them. (their shame, their cruelty,
even. The very things you were made to do in the past.)
Breach for a breach
Accident from the launching site.
A rocket exploded.
Puzzles scientists
New colors acidic. 20? As fingerprints.
Sexual behaviour. Heart.
To make them want to or paralyze.
In a way easy to hide
More difficult to find
You: brain 90°
Brain. Outside of.
As if you were two persons. Physical
The make believe. They just try to.
The knob: also for space traveling.
From an island to an island
We have (deliberately) caused civil
outbreak in Iraq, Lebanon, Palestine
Martyrdom and the macromedia
Committing will
Nations of the world as dominions
within the United imperial.
Iraq: Few hundred rebels captured. Their responsibility:

the university bombings
A market. A mosque.
Jazeera link
Soldiers. Their morale, iyw.
Drugs, Sex-abuse, treatm. of pow.
Who to be led etc.
Our rivals knew about our building site
Knew what we were building.
You. Ins. Equipm. To control the atmosphere better. (us in fact)
Like any a Versailles treaty: impossible to uphold
Anglo-Saxons kindle the oven, as before.
Superlense
Refractive index
Miniaturized antennae
Luther: doctrine of the two kingdoms
Law and Gospel.
Guru prophecy out. Economy control.
Arab youth run through
Peroxide (a chemical)
It somehow got loose
It just drifted
A mighty explosion
Containers, tanks. Big ones
Into the pacific
You: The works
Mostly women Indonesia. Gen, heart . . .
A highly active compound in head
Attracts from all directions
With great force.
Enter Gas,
Light.
Exit device.

Feel the skin in yr face vibrate
Tracts, cavities, carved lines, skull, brain,
Filled up every so often.
Lines of force within.
Hands, feet
Sight. Muscular harmony.

Jan 29th 07

The locust took a flight
Sophisticated addicts,
´heroic´ suicidals alike.
Lines of force conn. To glands in head. Ins: liquid fluid
We see them anew. They were practically invisible.
People driven through.
Within some Ghosts, siding
with the enemy. Shed doubt here, oppose there.
As if they left out a brand. (how to explain:
Photoluminescence, iyw)
A tiny movement
Expressing longings in the physical:
Hard to notice, if disciplined.
Quite a few sent to the ´darksome atmosphere´,
as T. Aquinas would say.
Diodes.
Lasers
  Pulsed
  Continuous
On the threshold:
Breakthrough in genetics:
Read the messenger. (forms + colors)
New elans.

More than a year ago:
Ins. In bloodstream,
heart. Compounds, essence from us:
To increase your population.
You´re less for us, and vice versa
Something similar now. Some of us
just have to see from here.
Don´t participate so much.
A year ago: space suit (on you).
Light-points on nerve-ends
Changing suit.
You have practically faded out
A few times these last weeks.
We´re the aggressors. We sure
Want to keep it like that.
Prosecutor Ghost.
Freud:New introductory lectures, (vol 2, penguin ´73, transl: Strachey)
1 Disposing of masturbation
2 liberation
3 suppressed although interest persists
as a defence against temptation.
M. Foucault. History of sexuality, (penguin ´78, Tr: Hurley, p. 77-83)
"Power-desire
Power as constitutive of desire:
You are always-already trapped.
Power prescribes an order for sex that operates
As form of intelligibility.
Do not appear if you
do not want to disappear.
Binary interplay:

Drive-consciousness
Illicit-licit"
Few things worse than addicted demagogs.
Buying pig in a poke.
The men interconnected.
Gas-cloud
Nietzsche: The Anti-Christ, Transl: H.L. Mencken.
"1# The storing up of powers
Christianity as degenerate rottenness, depressant.
The tonic passions,
Instincts suppressed.
10# Protestantism:
Hemiplegic paralysis of Christianity
16# national gods are either the will to power
or contrary to human inclination."
The doctrine of fascism.
Benito Mussolini, 1929
"The man of fascism is an individual
Who is nation and fatherland, the moral law:
Through the denial of himself, through the sacrifice
Of his own private interests, through death"
"Fascism is a superior law with an objective will
Fascism is a system of government
Fascism is a system of thought
Fascism overcomes the antithesis
Of monarchy and republic.
Activism: nationalism, futurism, fascism.
The state makes concrete political
Juridical and economic organisations
And is as such a manifestation of the spirit.
An empire. A territorial, military and mercantile
expression; spiritual."

Mein Kampf, 1924.
ChII
"The unholy alliance of the young Reich and the Austrian sham state
contained the germ of the subsequent world war.
Uncleanliness: moral stains on this ´chosen people´
like a maggot in a rotting body; spiritual pestilence.
Infecting people, scribblers who poison men´s souls
. . . the Jews, in tremendous numbers seemed
chosen by Nature for this shameful calling . . .
the relation of the Jews to prostitution, white-slave traffic:
a revolting vice traffic.
A long soul struggle had reached this conclusion. The great masses could be saved with the gravest sacrifice.
The only remaining question:
Whether the result of their action
in its ultimate form had existed
in the mind´s eye of the creators,
or whether they were the victims of an error.
The original founders of this plague of the nations
must have been veritable devils – for only in the brain
of a monster –not that of man-, whose activity must
ultimately result in the collapse of human civilization
and the consequent devastation of the world."
Hitler´s table talk (as recorded by Martin Bormann, (´41)
"Liberalism. Man´s mastery of nature and dominion over space.
Christianity: the heaviest blow on humanity. Bolshevism its illegitimate child. Both inventions of Jews.
In the line of power storm.
A deliberate lie
from the ancient instinctive respect and tolerance
as opposed to Christianity´s keynote: intolerance.

A scientific spirit.
The dogma of Christianity gets worn away before
the advances of science.
Religion is in perpetual conflict
with the spirit of free research.
Sublime suffering
a marked fanaticism a dominant feature of his spirit.
The Pentateuch became for ages the absolute rule
of the national mind."
Hitler shares Nietzsche's Verachtung towards Christianity.

GMT:0940

Feb 6th 2007

Over England
Explosion
: they will be severely punished.
Orbit polar point
From China.
From fear to anger.
Point shot down.
Counterattack.
Some ´former´ allies did not help
Germany, Sweden
Old communist countries,
Catholic countries
Loss:
Ten thousand
China: Few hundred thousand.
Russia:
Surrender.
They show us their men.

One of them made in haste.
Construction sites.
We gave them some material/tech.
They had a nuclear program
Accidents occurred moments afterwards:
So many on earth lost parts of their feelings, sight, memory.
(when we die, man loses light.)
Keys ; Rn5, P63, Lr7, Fm4...
2Dm>3Dm>point
Fold space
We had run 300 million through.
Some are just dying. Others in terrible pain.
Will be done differently.
Massacre
Murders
Martyrdom.
Prevented.
My merry Sherlocks.
That was the last episode of the cold war.
One of us knew. His warning signs passed unnoticed.
Structurally sound

Feb 7th 07

They think about fighting back around the clock
Backed up by their masse populaire
To cause unamnity amongst them.
Their most adjective (pervasive, wicked . . .) thought and
Their most hideous deeds.
To put UN in Afghanistan. Bad move
We nearly lost you.
Enter conscience Elans. Russia, China

Grouped together
Synchronize
They might give us a similar blow in the future, but we are sure
Going to prevent that.
Busloads of prisoners. Unconditional surrender
Constraint. Negation.
Historicism: history, as if correcting itself.
Old principates in Europe.
The doges of Venice
Our soldiers of all times, old fountain springtime blues
Old C- Europe
Royal houses
The Jews, as of now, deprived of fatherland; their necklace in the fountain.
The houses of Europe, colonies, and the new world.
Old unpolished windows,
Unattended wells,
        Things like that.

Feb 10th 07

Transition time
Castration. Thousands. Far east
Power brokers
Sweeping powers
Australia, Tasmania, N-Z. Strange communities.
One mind in the hearts of a million
Mansjuria. Shanghai Area.
The coast-line
Mongolia,
        Nomads
Tried to give you some sense of awareness.

Feb 16th 07

New Elans
Bones crushed.
We cannot hold Arabia.
Too many enemies
Weapons all too easy to come by
The lion´s share. A profit of a splendid isolation
USA mapped as enemy
Eternity correcting herself
Two youth revolutions in China. Suppressed.
We practically imprison their ghosts, magnified alike.

Mars 27th 07

A landing place in the middle east
3km radius
People from wherever died.
A young Arabic woman managed to
smuggle weapons in.
Knew where to shoot.
Inside help
Later. Youth Arabic imprisoned
Erikson, Erik
Childhood and society
Triad/paladin ´77 (55)
198 autocosmic play:
Perceptions
Sensations
Intellect?
Microsphere: thing-world
Macrosphere (world shared)

Autosphere
If frightened or disappointed in the microsphere: regress into the Autosphere; daydreaming, thumb sucking, masturbating
>see Freud´s liberation
375Human data: an eternal temptation to treat human data as if the being were an animal, a machine, a statistical item.
377 The inhumanity of colossal machine organisations . . . the danger emanating
from a ´total war machine´. Super organisation and cultural relativity endanger people who are in marginal position: it causes anxiety:
The basic virtues, (see p. 247)
Maternal quality of care and order.
Leibniz, Reihe, Politische schriften
La prusse ducale
Les princes en Europe doivent observer pour s´empecher, qu´il ne se forme
Une Monarchie Universelle (1685)
 -der keiserlichen politik
. . . tous les voisins doivent conspirer contre celui qu´ils tiennent entreprendre une
Guerre injuste, pour l´en faire repentir.

April 26th 07

Putin, (in his ´State of the nation´ speech):
"There are those who use pseudo-democratic rhetoric,
some loot the country´s national riches,
rob the people and the state,
others strip us of economic and political independence"
Any a nation could share these words with Mr. Putin,
I suppose.

Mai 3 07

Before. Early April. Iran put up shields.
They suddenly mushroomed up. To work their men.
Shot down.
Japan. Sent up an object. Had to respond. A direct threat.
Not to build arms.
You: people is so light within. They give themselves away.
Your aura. Movement within. (shape space)
We just see what the ghosts of the world are doing, and interfere.
Try to influence them.
Doctors with the knowhow. When the knowledge will be passed to individuals who are not within
From Newton´s principia

a a1 a2 a3 a4 a5
b b1 b2 b3 b4
c c1 c2 c3
d d1 d2
e e1
f

The Koran. Extr. From ´The family of Amrans´, 3:45
The angels said to Mary: Allah bids you rejoice in a word from him. His name is the Messiah, Jesus, son of Mary . . .
He shall be favoured by Allah.
He needs only say “be”, and it is.

Mai 31st 07

New Elans, E&M (entelects, magnified)
A comprehensive list: the men
So that every ghost be within

Experiments all over: to create new life form
Sth under the ice-cap
New software
Irrigating the sea
An equipment. When activated: as a mesh over areas
New space-craft
Turbulence in the ionosphere
Late 16th c. the English attacked a Spanish frigate returning w. gold from the new world. Buckingham loot. Proof established.
Old royal houses. How they treated their ′pieces′.
How they treated ′primitives′.
Disputes and quarrels
Shortcomings and ambiguity
Sth ancient. Krishna. Buddhas.
Deeds of the past. Eternity bleeding.
Age-old history revealed.
Opposing truths
Negligence
Early this winter: tech failure.
Great loss. Argentina.
Corroding enemies
British telecom system.
Synchronize
Unification theory
Adventures, fairies
as if rewriting
Cinderella,
Pinocchio,
the little red hen
Kerouak. Lost his beat.
Shirinovsky. His magnified and individuals within now.
Extreme right conn. Mob conn.

To verify sexual behaviour of all times.
New devices
You´re the core of it
Magnanimous task
Some old ghosts
Vespasianus
Apuleius
So much we can´t tell you. The secrecy of it all.
To have and hold, remember
Prism of detectors
Nuance in the atmosphere
Mass graves. Serbia.
On a Russian soil
Chernobyl
A chemical that ´eats´ radiation
Nuclear power plants. The worst of inventions
In the States.
Northerly
Leak.
Radiation.
Polar attraction
radioactivity around the north pole
Exploring the oceans
Deep down-under.
The world subjugated by one nation
The crown of lords
Neutronicity
Increasingly wireless
To rest awake for dangers
Rearranging the hierarchy
Under the watchful eye
The tree of life

The desicions-making
How to interfere,
influence,
intervene.
Conqueror-Ghost

June 3 07

Deflation time?
Rommel: Krieg ohne Hass
Russia attacked us
Counterattack
Thought done anew
Analytical chemistry
Spectro-EC tech
Photochemistry
electrochemistry
Properties
Photovolt
electrochrom
Light producing reaction
Deprotonation
S/N: signals-to-noise
Tuberculosis: blood
Breast cancer
Oval: 1/100 woman deed
Fertilize
Human waste in the Baltic
Marat. An English dagger.
Bakunin, M. God and the State, 1871. (see fx Marxist archives)
Explaining to the profane
Emancipated from animality

´by an act of disobedience and science: rebellion and thought´
-
Bucharin, N. Poetry, poetics. 1934
W. extreme sharpness: the problem of quality is diversity
The spirit, like transmogrifier
The Great Tao lives in a state of inexpressible
poetic inspiration. (Bakunin on the Chinese poet,
Kiun Tu (+909AD) the harmonious loom).
ChI: idealistical or mystical points of view are unacceptable.
The process of life has its intellectual, emotional and volitional sides.
Thought, concepts and images make up the realm of emotion.
Dialectical magnitudes s.a. feeling-intellect
Conscious-unconscious
Composing unity
Thought
Type dialectics
Product science
X
Opposites, elements. Essence
Essence merging into the phenomenon
The ABC of Communism
(Bukarin and E. Preobrazhensky)
A programme drawn from life
#1-6 Refashion the whole world to suit themselves (the workers).
The old order is collapsing under our very eyes.
On Capitalism: commodity. Private ownership contains
the germs of large-scale ownership, the means of production,
the product itself.
The labour power becomes a commodity
Production of surplus value

flows into the pockets of the master class
Capital
Monetary, industry
Commodity form
Demarcation of energy
Insolubility of the complex nature habits: born wild
Habit forming
Excesses of a lifestyle
Universal catharsis
Karl Kautsky. The Class Struggle (1888)
Ch III Trade-unions and political activities
Common property: that the workers
appropriate the surplus.
A system of production for use (as opp. to ´for sale´). Free from the fluctuations of the market
As a sequel to the capitalist system of production: the transformation of the separate capitalist establishments into socialist institutions.
Engels: (1894. See fx Lenin 1917: The state and revolution: The 1891 preface to Marx´s "the civil war in France´).
The programme is downright communist.
A generation reared in new, free social conditions.
Bergson, Henry: *L´evolution créatrice*, 1907
P. 67: *élan original*
élan commun
un contraste entre complexité de l´organe
et l´unité de la fonciton déconcerte l´esprit.
Une theorie méchanistique,
la doctrine de la finalité
186: un principe de crëation.
L´existence m´apparait comme un conquëte sur le néant.
Existence logique

Psychologique
Physique

His dialectic (here) may be constructed in the form of:

E
C D
A B

Perpetualité – processus- x
Colloid and interface science
Emulsions Liquid in liquid
Dispersions solid in liquid
Aerosols/foams Liquid in gas
Genomics
Abstract information
Probe
Unlock
Provide

<C-terminal-binding protein interacting protein> (Nature June 4 07. Authors various., x)

Binding protein partners considered to contribute to transcriptional repression and cell cycle regulatory properties. A role in the cellular response to DNA damage.

Transactivate the gene reporter

Expression (CtIP) is induced during viral infection. Reduction with RNA interference.

Possible depression.

<A novel glioblastoma cancer gene therapy> (Auth. Various. Nature, June 07, x)

Transient expression over expression of the
C-terminal fragment
of the human telomerase reverse transcriptase (hTERTC27) exerting
immune responses

signal transduction
transport transduction
<miRNA control gene expression in the single-cell alga>
(Nature. Various. Email: david.baulcombe@tsl.ac.uk
MicroRNAs in eukaryotes guide posttranscriptional regulation by means of
a targeted RNA degradation and translational arrest. Released inverted repeat forms. Recruited by the effector protein of the silencing mechanism short interfering RNA (siRNA) released/produced. They guide post transcriptional regulation, as with miRNA, and epigenetic genome modification.
(Zamore, p) siRNAs trigger an immune response against
viruses by destroying complementary RNA sequences.
<Cultivation of a novel cold-adapted nitrite oxidizing
Betaproteobacterium from the Siberian arctic> (spieck@mikrobiologie.uni-hamburg.de>
Permafrost affected soils investigated. Prevalent cells w.a conspicuous ultrastructure
a novel chemilith germs in polygonal tundra soils.
-
<A polymorphism in TLR2 (toll-like receptor2)> (Nature, June 07 sdunstan@oucru.org).
Tuberculosis meningitis (TBM) results from dissemination
of a bacterium from the lung to the brain
Racial diversity.
Compare:
<Microbial communities found in healthy Caucasian and Black women>
The maintenance of low pH in the vagina through the microbial production of lactic acid as a defence against infectious diseases.
Mechanism. 8 major kinds of vaginal communities (super

groups) studied, based
on the profiles of terminal restriction fragments of 16SrRNA gene sequences,
not all super groups equally resilient, which account for disparities in the susceptibility and sexually transmitted diseases.
<Genome-wide association study> (Nature 447 June 7th 07, donnelly@stats.ox.ac.uk>
The GWA represents a powerful approach to the identification of genes involved
in common human diseases. Case control comparisons. On the basis of findings and replication studies, these signals reflect genuine susceptibility effects. Associations at
identified locii. Shared control group: a Data, result and software for exploring the pathophysiology of disorders.
Phenotypes – genotypes- x
Characteristics of organs – genetic constitution -
Chlorophylls
<Nuclear microenvironments in biological control and cancer>
Authors: various.
Nucleic acids and regulating proteins are compartmentalized in microenvironments within the nucleus.
This sub nuclear organization may support convergence and integration of physiological signals for the combinatorial control of gene expression, DNA replication and repair. Nuclear organization is modified in many cancers.
There are cancer-related.
Changes in the composition, organization and assembly of regulatory complexes at intranuclear sites. Mechanistic insights into the temporal and spatial organization of machinery for
gene expression within the nucleus, which is compromised in tumors,provide a novel platform for diagnosis and therapy.
Building blocks hijacked

Large structures rearranged
Holographically controlled optical traps . . .
Bacteria, parasites, viruses
Cyg X-3
Quasars, 3C 273
SN 1987A
Progenitors
Periastron separation
W.machine precision
Discretization. Numerical approximation
Defining elements by shape functions
Region of support
Elemental degrees of freedom
Rotational displacement
Element matrix
Nodal frequency in the mesh. Very high
frequency response. Material n-n linearity. The mechanics of an operation search parameter.
<Finite element analysis 1999 Laursen, Attaway, Zadoks. www. osti.gov Osti ID4711>
P. 42For any proposed element formulation. Patch test: boundary value problem imposed on a patch of elements in constant strain (and thus stress) solution, and to demand exact numerical solution.
Systematic performance.
Continuum mechanics.
Time dependent external.
Coulomb field
P. 7 core collapse supernovae. The core collapses due to force of gravity, and stops when the nuclei form a dense phase of nuclear matter. The pressure increases, pressure waves propagate outwards.

Sonic point: velocity of infalling materials exceeds the velocity of sound.
Leads to a shock wave: collapse and bounce.
For the star to explode: a shock reheating/reenergizing mechanism.
Integral part a convection
models
<Microscopic heavy –ion theory> Ernst, Oberacker, Umar (1996, OSTI ID 604344)
Muon-induced fission
K+ > center of Ca nucleus
While the high energy pion will penetrate
to the center of a C12 nucleus.
Properties of neutron-rich nuclei
Quark-gluon plasma phase to deconfine the quarks
and gluons of baryons and mesons, Lepton pairs (EM) produced from hadronic interactions for direct information on space-time region
Feynman diagrams
Dirac equation
EP pair production with capture w(U92+ beams at a KE of 01Gev per nucleon.
Compton wavelengths
Galactic poles
Galaxy modeling
Dust and synchrotron contamination. Radiation
<The CMBR spectrum> (Author: Albert Stebbins, Fermilab 1997, OSTI ID 646367)
Doppler/gravitational redshift.
Brightness: specific intensity of light.
Spectral distortions in terms of brightness temperature versus the dimensionless frequency.

June 15th 07

All kinds of magnets
Authoritative within a renaissance Christ. Used Christ to hide their deeds.
Eternity as an organic whole. A contrary act, surprise, mistake, . . . can cause
destiny, fortune, fate, to intervene.
You´re the axiom through which all of us have to go
The four great rivers of faith come together within

June 18th 07

Biomarkers
The German WWII Ghost: the Jewish men were free to leave their soil. Freud, Einstein,
Recreation of the past. WW crimes.
Guilt-innocence
Melatonin
colonies within, (sort of in vitro).
Auras. Marked by deeds
In the void. Photonic blast. Experimental
On the space project:
Sun-light (prblm)
Solar panels
New strand of D/RNA discovered
In mammals: the moment of dying
in human beings: the hour of departure
The Buddha of the latter day: Nichiren Daishonin, (1222-1282),
The casting of the transient and revealing the true identity.
Extracts from ´The universal salty taste´ (see fx SGi library

(soka gakkai)
The Ocean and its mysterious qualities. Its bottom is hard to fathom.
Means metaphorically that the realm of the Lotus Sutra can only be understood and shared between Buddhas.
Creatures of great size exist and dwell in it: The Buddhas and bodhisattvas possess great wisdom, they are called ´creatures of great size´, and that their great bodies, great aspiring minds, great distinguishing features, great evil conquering force, great preaching, great authority, great transcendental powers, great compassion and great pity all arise naturally from the Lotus Sutra.
Soka gakkai, Its ideals and tradition, by Daisaku Ikeda, 1979:
P. 12.: The key to all of Sakyamunïs (Siddharta/Gautama) teachings is the Lotus Sutra. The purpose of Lord Buddha's advent in this world lays in his behaviour as a human being.
P. 11:one of the goals of the Soka Gakkai is to guard to the death the freedom of the human spirit.
P. 15.According to the Vimalakirti Sutra, (Vimalakirti is a bodhisattva of Mahayana Buddhism),word at one time reached Shakyamuni that Vimalakirti had fallen ill. Disciples were urged to go to Vimalakirti´s home and inquire about his condition. When asked, Vimalakirti replied: "because all beings are sick, so the bodhisattva is sick. If all beings recover, then bodhisattva will recover. The sickness of the bodhisattva arises from his great compassion."
Geobotany
Geotropism
Gene flow; exchange within populations
Med.rep:
Guts
Kidneys. Excretion. Regulation
Gastrulation: Endo-meso-ectoderm

Urethra: string inside the duct
Urinary bladder
Digestive organs
Taste. The tongue. Nose.
Mouth>ears
Throat
Teeth
Jaw
The young going through
Zygote: a fertilized ovum (before clearage).
Contraction cannot be imitated as well as we would
have wanted
Acidity-alkalinity

June 22nd 07

Liver>bile>gallbladder>duodenum
Irritability
A universal property of living things
Complex, adaptive activity
See fx ´Biology´, by Abercrombie, Hickman, Johnson, 1951:
Pancreas (situated in mesentary, near duodenum) discharges through a duct an alkaline mixture of digestive enzymes, (trypsynogen, lipase, amylase…), an organic substance stimulated by hormone secreting (from gland cells to bloodstream)
Also: secrets the hormones, (affects the amount of glucose in blood. A protein. Suppresses breakdown of liver glycogen from glucose. Antagonized by adrenaline, glucorticoids, stimulates protein glucagon synthesis). P. 149: Insulin, glucagon . . . acc to the level of blood sugar. (Supplied by the liver).
Glucose, as glycogen, removed from blood by all body cells as food. If too low, (hypoglycaemia),: damages brain cells. If too

high (hyperglycaemia): diabetes mellitus.
Glucose excreted by kidney
Hormones interacting
Phosphate
Fructose
Endochrine
Hormone secretion
Proteins

June 23 07

Conspiracy. 20 nations participated
Saw it coming culminated in attack from Russia
Answered fiercely. Had given them some tech before.
Won´t be done for possible enemies.
Parts of ghosts in the States
Part of Ghost Portugal>S-Am
8oo Russians
Quite a few from us
They killed some supposed moles
Some time before. New telescope. ´eyes´ everywhere.
Built. A shield for Israel. Has been rammed.
West coast pacific defence wall
Tungsten, Vanadium
Exp. Within
Every metal w. its own character
bombarding
Radiating energy
Force field
Chemical reaction.
Regrouping EMI
(entelects, magnified, individuals) within

Deceit. Illusions
Few new elans
Found: old sibling sin all over.
Specks (sentients) from nearly every individual on this planet in our hands.
(within the economic muscle). As if on a plate.

June 29th 07

The Japanese found it.
Neutron ventilator. Seen.
Experimenting.
Pre-emptives
Have a torpedo. Long range
A device to track our ´eyes´.
Know about our ice-cap
Arabia. Were preparing sth.
Response: Photonic. Thousands.
Pre-emptive.
So that this war wont culminate on ground
Military on the move. Prevented (persuasion) few times.
They could annihilate us within a minute.
Losing allies
Supercomputers
High performance computing
No innate ideas exist
A pearly drug
Leaves traces
Start the generator
-the shields are up
Activate yr photonic
Demolish the power

July 4th 07

Over English soil
Exp. W. fuel reserve for
Space project.
Accident.
Molybdenum
Fingers, toes
Arabia. One of their men
Breeding suicidals
They just love them till death
Our children.
The lines of movement
If they are in pain
They just feel for their insides.
Break a leg
Sayings. Proverbs. The thought behind
Inside the plasma (means something else than before).
No body. No skeleton. Molecules, elements, fundamentals in
Constant motion.
As has been noted: your satisfaction. As anything else.
Nothing to you. Makes you special, to say the least.
Satisfaction makes people feel good.
The ´come´ and emotions accordingly. Emotionally attached to.
The sweetness of it. Relaxation. The lot.
The densest material. A filament, 1 micro thickness.
In regions in the brain.. Machines and plasma in and around
The physical. Made to come. Proof of burden:
The penis shouldn´t go into the mouth. Unclean.
The tongue; vagina/or anus. Punished alike.
Small devices inside the bones, (various places), for light of movement.

Aureal lines.
From center of brain (pit,pin,th,)
Rib, shoulder
The world made through your heart.
Thousands at a time. The women. The work.
Quite a few young men. Break the penis. Proof.
Light of movement. Aureol.
W have to be the physical
A physical extension.
-(we left out a brand. Other reasons too)
A kind of a key. They just can´t any more.
We give them to ´come´ then.
Fill you up
Liquid in and around you.
Gas in brain
Activate
Quite a few
From each continent/country
To give them the feel for it.
Recovery time.
<the same again>
New elans
Yr heart so thirsty. Certified.
Physical release
In a gas cloud
Controlled, (edit, punishment), by strength
Food (glnr prdct) fulfilment of longings
The loving of it
Movement
Size.
Sri Cinmoy:
From 'My flute'

"My name is god´s eternal game
Another day, another day "...

July 8th 07

Burma
Pol Pot.   Root OBL within
            Man Philippines
What happened at Kakimuni?
Someone found out how to hide a deed
In the eve of witchcraft. Thought his friends.
Has been spreading up until now.
The skin. (the flaying of it)
Through mought
Ins. Skull
Out neck
Inside back (cortex, point hands)
Device
Women deeds
The pussies of Christ, Buddha, Krishna, Allah . . . .
Masters of illusion

July 11th 07

U.S.A. lies
Japan
War in between islands
Wanted to help. Accident.
It exploded. Few hundred died.
Sth. In the national
(English) character
The Boston-Strangler found
The ambassadors of music:

L. Armstrong
Elvis
Lennon
David Bowie
Elton John
Pavarotti
B. Marley
Hobbes
Leibniz
Kant
Hegel
J. Bentham
Che Guevera
W.R.Hearts
Oppenheimer
Lawrence of Arabia
O. Bin Laden
Saddam Hussein
Arafat

July 13th 07

Arabs. Attacked a Christian family
Threw explosives into their home.
Before: Arafat. Tried sth.
Cuba War. On the verge.
U.S.A. Schwarzenegger> U.S. Austria man,
Activity in Austria.

July 15th 07

Singapore Mahayana (indoch pen, Indonesia, Java, (sufists))

Spain
Arabic penins.
English
Japan
India
Sri lanka (Tamil)

France
Sarchozy: activity in the S-E. Russia dialogue.
Root. Mendela. Down. Part Churchill.
Putin-Chaves
Smuggling explosives
moles within.
Iran. Achmadinachad, Imam Chomeni: to get to Israel
Go through England
S-Amerika:
South of Sahara. Activity
Indonesia(Java). Activity.
Castro.
Another man Cuba.

The leaders oppressed

Before: The English-Japanese betrayed. (Lennon).
The greatest loss: Africa.
Pre-emptives, aid.
-(Please forgive us for this mortal sin).
Melatonin
From ´The manifesto of Umkhanto we Sizwe'
´The spear of the nation, Des ´61 see fx wiki:
The time comes in the life of any nation to submit or fight . . .
We have no choice but to hit back by all means . . . The methods of Umkhanto mark a break with the past . . . . to bring the government
And its supporters to their senses.

Chief Lutuli: Let our courage rise w. danger.
N. Mandela: No one nation should be a complainant, prosecutor and judge.
Yr right arm. The dermis and the lot. Flattened out. Fibers, tissues.
To recreate the tracks from the bladder to the vagina, the oval system.
(the very hands)
Colour proof. Soft pink, w. a dint of yellow/brown.
Why some women don´t get satisfaction. Our young girls.
To urinate.
How we touch. The wanting of it.
The great homecoming.
Yr. hands. The skin, tissues, opened up.
Heart diseases. See Fx.nlm.nih.gov/midlineplus
Endocarditis
        Bacterial infection
Cardiogenic shock
        Disorders of the heart muscle, valves or the
        Electrical conduction system
        Symptoms/signs: low blood pressure
Heart attack. (Myocardial infarction)
        Oxygen loss. Heart muscle
                Dies or damages
        Blood clots
                =atherosclerosis
        S: diabetes
            High cholesterol
            Fat
            Obesity
Acute/chronic mitral valve regurgitation
        Eg: the mitral valve does not close properly.
Breast Cancer

(most common)
Ductal carcimona: in the tubes
From lobules to the nipple
Labular carcimona: in the lobules.
(See fx H.Gen. prjct)
A mother cell. Shortage or excesses. Transition
Causes chain reaction.
See fx: Geonomics.energy.gov
Hunter,T.: Signaling 2000-and beyond
:Signal transduction/amplification
Modulators
Pathways
Switches
Signaling by direct intervention
Receptors
Activate protein kinesis.

Signaling proteins
Effector proteins.
Binding domains
Membrane micro-domains
Activation loop
Feedback loop
X
Homologic
Homotypic
X
Inhibit, block or ban
Catalytic activity
Lipids. The second messenger.
EST-databases
Enzymes as targets

Widespread involvement of
Protein kinesis and phosphates in disease
Drugs to activate or inhibit.
Receptor – carrier – transmitter
A protein essential
For a given response.
Autoinhibitory mechanism.
=Cell, Vol 100, pp.113-127
The three ontologies of a protein. (see fx: geneontology,org)
Process
Components
Function

July 23 07

Eternity done anew.
Abdomen
Inner organs outside of
For proof coloric, eliminate movement,
Sedative, fluid in brain
Skin off
Skeleton off into the head, out anus, point objective
Nodes within
Bypass organs
No intermediary for the 'come'
Histamine
Either feed them (w.glnr product) -or deprive them of-
Individuals are too strong within people.
Newcomers will be different..
Thrd suspect Visco. Confessed
One of the first to work within.
The things you were deprived of

forms national, character.
One of our astronauts died.
Millions of Km away
The pressure dropped suddenly.
Both knees
Wake up in thin air 1000x
Heart carved
The world all over.
Specks of light
Circulating at a great
Speed around a center,
Within the flesh
Touching organs
Distinct places
:job done in brain.
Individuals were practically
Home free as second generation within.
Proof of burden.

July 27th 07

Arabia
Gaza. A kind of a plastic explosive
Oxygen driven
Japan
North vs. South
He-ni? Wpn
Night. Gaza up in flames.
Rocket into Israel
Fierce response
Lawrence of Arabia, Churchill, others. Lost some of their
Branches in Arabia. They just killed them.

Elvis in Russia.
Part Lennon in Japan
Purification not expected.
Russia-China
Industrial parts
Transp. Across the border
Wpns prd. Found
Arafat, other. Their young carry the messages.
The muscles cut loose one by one
The aura beyond geometrical boundaries
90° turn easy
Here: trading places. The lot reincarnated, iyw.
Follow yr own karma. (Chosen of course. By deeds committed, talent, taste, age,)
A former addict must help someone else to quit.
Things like that
In a way present in the one before
Guardians of youth, memory, feelings.
Wake up as an adventurous hare, a mad hatter
Or as the queen of hearts, at will.
The classroom dilemma:
The teacher focuses (without even intending to)
On distinct individuals according to
his methods, taste, likings.
Here: few fatal accidents from within
Frontal lobe, mouth, tongue, teeth, ears.
Tremendous pain.
New prisoners.
Sentients through
New locii (wpns, activity)
Find the engineers, doctors, the general ghost.
C. Stevens. (a part of him knew TT)

July 27th 07

Next one up
To bring the heart outside of
Nerve ends filtered w/light > light outside of . Devices.
Ox from blood in brain
Yr dizziness
Our instruments, devices, chemicals eat it.
Always low
Hamas Kandahar man
Gaddafi
Arafat
Venezuela
From the inventive site
New (aura) marker
Acidic color
Yellow-white-red-blue/ white magenta yellow
Food: thought
Mixed feelings
:to undo thoughts.
Find
Make them think faster/not to think..
The bath. Within. Magnetic field. Gases. Biomarkers
Women from wherever.
The marking of the aura
Yr lines of movement, when within: split up, 20x2x
Animal symptoms. Fertility in domestic animals
We just hurt them.
Ancient Egypt
China: preparation
Response.
Russia. Missile attack. (Rammed)

Response. Where there was activity.
Petersburg-man.
Japan. Southern-islands-man
N-Jap
S-korea
Indiv. Within:Shogun, Gangmembers
Addicts, Prostitutes, ww2man: The lot
Very nationalistic, magnified
Teargas harms the eye-lids

Aug 1st 07

New elans
S-Jap Eentelect
Africa 3 entelects
The assembly
Togetherness
Literally every sovereign country
Only ten nations still
Fighting against it, holding
onto some constitutional right,
written or unwritten laws:
to defend their country. Other reasons:
conqueror-spirits, (never give up); autonomy
of the nation at loss, ignorance.
Rearranged by continent
Interest
Need
Stomach, duodenum, rectum; abdomen
Defecation habits around and within.
For months now.
Laziness, teaching the young, etc.

Nourishment.
New elans
Ruanda
Zimbabwe: R. Mugabe
Afrika: Raid. Tens of Thousands. Traffickers,
Addicts, murderers, rapists, . . .
Civil outbreak possible.
The neglected continent. Why? The world wars, fx.
The colonial borderlines
The continent has to evolve geographically, demographically
Mass graves
People killed in prisons via individuals within.
The economic system in ruins.
Just corrupt from the tip to the toe.
They didn´t even tag/mark people with AIDS, let alone
Prevent them from intercourse.
We lost faith in you for a while.
You lost more than you came for:
The visco-incident.
North of Sahara.
The gardens of Sativas
>Falangists
>Netherlands
Belgium. Sexists. Highly vulnerable
at a point in time. Have had to carry that load.
Ketamin
Fermon
English forte: studied insult
Curiosities: Wilhelm Reich, Joseph Smith
Russia. Counter espionage
Chernomyrdin
New surveillance system

Software
Persuasion. Moments after turned off.
More flaying of the skin
Left foot.
The skin, tissues, nerves cut off. Conn to gen.
Old tech
Laser, wavelength det. (radar) still in use Indoch. Peninsula
Colombia. A bit like R2D2 (without the intelligence)
Economics. Securities control.
Thorough investigation
Corporations, companies, firms
The Suisse banking system
More transparency
The Alps. Avalanche caused by us.
More than a year ago
A violent move. Few hundred within died.
Man Iran
Bangkok
Bangladesh
Corroding energy
The foetus
Some genetic/chromosome defects after conception.
Some types of cancer
Diabetes
Cleft palate
Downs syndrome
Some symptoms that break through later in life: not as apt to do sth.

Aug 8th 07

Burma. Just see their Shrine. A common sin traced to them.
Uruguay

Indians, slaves
Micro-lasers
Gluon force
In one of the States.
Construction
Not aimed at us, was said
Prevented none the less
Strong blue
Red bad combination
yellow
Marley: indiv. Traced to Colombia
Coca leaves
>Azor (negro society)
>Miami
Venezuela
Hugo Chaves
Neighbours
Russian
Che Guevara >Castro
Drug related, women marked
The women, so many of them move their hands in
A very distinct manner.
Russia
Attack
Shot a Site down.
Counterattack
Gas poisoning
Gas-line out. Resources.
Indiv. On earth in danger.
Casualties.
Here: amphitheatre. Addicts.
One of them died.

One of the joints came apart.
We were running them through yr brain.
Yr countrymen sent a film from the ´split´ to the Haague.
Were experimented upon.
The cruelty. The lies.
Feed and lead. Never falter.
Venezuela again. A few killed themselves from the root.
Which means of course that they are within someone
we haven´t done, or escaped our notice.
Those marked are put on a kind of a photocopic tape.
Grain of material in heart, heated, turns blue. Elans put into the brain. (others in the heart instead).
A sparkling new technique. Machines, devices, software, hardware,
Fuel and manpower.
Elans rearranged
Man Colombia. Farms.

Aug 8th 07

Accident in the States.
Like a raging storm
Casualties. 200.000
Intoxicating democracy
Wrong decisions-making at that time.
In the heat of times
A Ghost was working on earth.
Died. A woman magnified. The same.
A great national loss.
Some international affairs too.
Your countrymen of all times
(mostly from Ghosts Norway, Denmark,)

Grouped together.
Vivified within
Historical standpoint
A material we use in industry.
(fission) Doesn´t reproduce.
Reserve: 50 years. The factories have to close down.
Vital. The atmosphere will collapse
Without this material/compound.
Rearranging, regrouping
Serotonin

Aug 10th 07

Respiratory system
A node within: urinate.
    To recreate yr nodes
    As if a woman´s
Blood circulation
The points in the brain that ´are´ the brain
300 millions through. 1/50000 dies
To eliminate movement. Abdomen.
Face carved
The back. For movement
Work done in brain
Distinct physical parts scanned.
One Ghost. Sixth generation. Died within.
One region in China.
Too many young died.
A movement that had not been listed
N-Amerika
The Ghost was working in one of their men
Here: middle aged loneliness crisis. To synchronize.

Manicheism: erroneous

Aug 11th 07

Hungary. Missiles from Russia. Energy source (English) down.
Few thousand from Hungary +
Enter man Petrograd. Within Koturbinsky
Has attacked us two times.
The Moscow theater incident
The school in Tschetsnia.
Proof established.
Extremely simple missile tech.
Machine-parts for other wpns.
System failure, (unaccounted for) We were too late to respond.
Someone within aimed. Forgot to lock. Missed at that time.
Another man Finland. Pro Russia
Enter man Central China.
    Killed his Buddha from himself.
Man Burma
Man Sudan
    Uganda
    Arabia
Man Pakistan
    Musharaf
Mesocompounds
Mesotechnics has evolved. Mesonics.

Aug 16th 07

Thought we could cure Chernobyl children. So hopeful.
One or both of the parents have been exposed to radiation.
The deformity inherits. The very soul takes on the same form.
Novgorod. Building site.

Down.
Arabs. Confrontation every day.
A doctor found.
A man in the making.
First thing they do is to throw arms
up from him.
Soviet. The strongest reason for her downfall:
Chernobyl. Their shame. The northernmost part
Of Ukraine. Belorussia. We can but to try to
understand. They kill their Chernobyl victims.
Willingly or not. They just can´t cope with it.
Their neighbours look away.
The world closes his eyes.
The organs are misplaced, misshaped.
Such deformity. Obviously not the same
As acquired symptoms/diseases.
Will infect generations to come
Hospitals. Badly equipped.

Aug 17th 07

Beyond the pressure point, new material forms.
New energy source. A gas reserve from the crust
Deep under the surface of the ocean.
Force field. And instrument or two over an area.
Won´t have them throwing things at us.
They couldn´t even invite them within. Couldn´t
Give them movement.
Showed us their pre-emptives. What they themselves
Had done for the victims of April ´86.
Few new colors.
Your organic structure rearranged.

Metal tracking device. (transitional)
The East. Russia. Attack. (system in submarines)
From Bangladesh. China.
China-USA man, (Locus USA), passed unnoticed.
Mayakovsky. His cloud. The roulette.
The one who pulled the trigger. Found.
Quite a few new individual elans, Belarussia.
Man Belarussia.
The young
Women
Aristocrats
Soldiers of all times.
Dunant (founder of the red cross)
(surprisingly neutral, don´t you think)
LSD factory found. Netherlands.
Heroin. From eastern Europe. Frozen. In a food container.

Aug 20th 07

Optical tracts.
(Wwhite man´s) burden becomes shame
Ghosts lose autonomy
Language machine.
Our language, although ingenious, (morphem, phonem, lexem).
Too whimsical.
Indochina. Deep in the jungle. Their man.
S-America
The eastern bloc.
Insider.
Pre-emptive.
E-Germany
Corpses polluting water reserve.

The Russians dumped them there WW2

Aug 30th 07

Iran. Building arms
New laser tech
Pre-emptive
Fluid within. Bath. Children
New elans
Man Argentina. (Maradona, resp)
When in Naples. Tried to use him. All too many addicts within.
Iranians. Preparing big time.
Achmadjinahad. Lost a branch
Someone else too.
South of the gulf
In the desert. A site.
Parts for missiles>Somalia>Sudan
Try to bypass our conscience elans.
Enter entelects
Japan. Activity on an Island in the pacific
Mom. (military on the move)
Suppression
Photonic. Casualties.
Russia. Threw up a lite
Persuasion
More than 2000 men alive to-day
Jan Mayen. A site. (Ours)
Off the coast of Alaska. (The same)
Tens of thousands of women
Didn´t take heed.
Run through. Can´t any more.

Insert: inside the vagina
Some bleeding to death
Slight adjustment
Will be done none the less
Israel. Gave them ′eyes′.
Good move they say.
Pakistan. Bombings
Explosives found
Perpetrators. Give them it.
Back of head
Drill
For great quantities.
“Overwatch” Brown, G
Sudan. (BBC)
Underscore
Sticking points

Sept 09th 07

Thought crystallizing
Intestines. Recreated. (throat, skin, tissues)
Semen, ovum. Fertility.
Hemorrhoids
Virus infection
>improvement of drugs.
Travolta.
His doctor within Hubbard. Founder of Scientology
The Elephant used
His cortex.
Fx, plates
Pavarotty, Luciano +
′be a happy ghost′.

Oct 9th 07

Ingutsea. Raid. Drugs. Few thousand. Traces
All throughout Europe. Tr. To S-Am.
Ever increasing activity in Russia. Wpns programme
Both phases.
Persuasive power. Response. Kill us from them.
Vessel down. We lost a few.
Their loss tremendous.
Dniepr polluted. Sth. Leaked into it.
Their Allies: Iran. (China, Jpn, Vietnam, Indonesia,)
Used gas compound we hadn´t found.
Comm sys netw. Down
Plants, factories.
Their Moscow ship. Corrupt.
Their magnified need to be done as well.
Their secret hide-out found. Lab.
Pascal. Had Russians within.
Compensation.
Enter man Indonesia

Oct 10th 07

Japan. Have the rules. (arithm)
Fooled our surveillance system. Highly secretive.
Experimenting w. a deadly. In a crater. (A week later wld have been too late)
Enter man Japan
Allies: Philippines, Burma, (provided them w. material)
You: aura outside of.
Enter. Their numbers into computers
Equipments confiscated. We just stole their tech.

They think differently. Found errors in our calculations.
Possibilities we hadn´t thought of.
More military activity on earth, in the pacific.
Suicidal on wrist. When activated it eats sth vital
In the one who bears it. Dead instant.
To secure peace
Londonderry. Few died. Wpns smgld.
Macedonia
Rumania

Oct 10th 07

Directly afterwards
Japanese individual
Run through
The world all over again
(how we use the aura: central brain, glands,)
Mining for hrs
Every second day
Milochevich
(M) Marco, Noriega, Tito
Brown, G
Bush, G
Hussein, S
Disarmament
Great casualties
Vietnam
Pre-emptives
Tripoli
Old Acropolis
Anaximander
Some men who were destined to-

But died young.
Accidents, something not foreseen . . .
C. Cobain.
Mozart, Lennon
M.L. King (political).
Elvis, Andreas.
Breshnef
Shostakovitch
An English woman within.
Show′d them what we were doing.
Had given her a chance before.
No war without a Mata Hari..
Two or more within.
One of them opens up (makes her feel that she wants to)
Her vagina. Another on stretches out. (for a rapist). The latter
Had been punished.
The aura: from one individual within to another.
The physical hurt.
New routes. More efficiency.
Transparency of mankind.
Horizontal/vertical dimensions
Some indiv. were crumbled within a ghost
Material + gas
Color/light.
Luminous intensity
Light. Interval at will
(M) Pinochet, Sharif
Musharaf

Sept 12th 07

Enter magnified Japan

Enter man Russia
Man Pakistan
Arab pen
Karadzik
Indonesia
Bangladesh
Sikkim
India
(Closed our windows first).
Have been closing in on him.
Many of the suicide bombings in Iraq.
The Pakistanis owe the Iraqi nation an apology.
Fooled us w. an apparent friendship.
Heroin from Pakistan. Cargo. Netherlands.
Japan. Experimenting w. the new tech.
All too easy to build radars, lites, hardly detectable.
The man Australia-Japan
Japan smuggled wnps to the mainland
Hide too well.
Know the possibilities of this new tech
Can see them differently now (from within)
Filters, chemical agents.
Russia Pakistan connection.
Cold war wars were fought on a third ground.
Maybe that was the scheme.
Peru
The Inks. Encounter from Asia (China).
In a distant past.

Sept 17th 07

Prison intact

Enter Conscience elans
N-Korea –man
Japan man
Russia man
Ukraine man. (indiv. from April ´86)
Every second day a confrontation w Japan, Russia, Iran,
Power bleak.
Ins. Gas, indiv. light, glanular product
The strangest ring
The interlude inbetween individuals.
Entelects, magnified, individuals, (EMI)
Within or in history.
Homosexuals, Lesbians. So unclean.

Oct 21st 07

Yr heart stopped for an instant
El shock
Bush. Left wing Colombians tried to kill him. Marginal effort.
S-Arabia. Small fast vessels (´+2000 km.pr.hr.) in the desert
Traces: Dubai. Iran.
The drop of life
The men who came in your wake.
Their sense/feeling had parted
As was common.
The drop of life. The doctor ghost knows how now.
Bypass. Create a bridge.
The men to come will be alive. Their own feelings.
Part pure thought though. Causes problems: pain,
character creation,
Milton:
Hear all ye Angels, progeny of light

Our spoils
Handicapped from wherever.
To give them light of movement,
Physical appearance, take away pain.
Cure blunt.
Always som part of your body under attack.
We make them believe you´re someone else.
The 1972 olympics.
Found. Within Arafat.
Musa Al-Sadr
Old traffickers/users within. (Opium). Light shed.
Desperate attempt to save one species of a whale.
3 individuals left. One female.
The Bengali Tiger.
Few thousand mammals on the verge of extinction. Ameophebia
Thousands of flowers, trees.
Blue fin tuna. In a good year: few millions.
In a bad year: few hundred thousand.
The tropical rainforest is just dying
Images from galleries long gone, paintings, mosaik . . . (wars, sabotage, incendie)
The library in Alexandria. Incendie suspicions.
Couldn´t find any link.
The sphinxes. Their faces known.
Antonius and Cleopatra. Political murder.
A pyramid. Hitherto hidden.
Locus known.
Hindu shrine. Indus valley
The cradle of eternity.
Patience and cunning
The cow. Symbolic significance.

Life perseverance
Pakistani killed their Krishna from themselves
By the time Muslim population expanded.
Mammary glands
Part of the thermal system.
Bad companies. Lose by command.
New market theory
India. Few millions through.
Lost a marker. Few thousand died.
Ukraine. Janukovitz
The Arabs. Sacrificed one of their men.
Muhammad Atta. They just died from him
When the flight ended. (mind me, quite a few of
The passengers are with us to-day).
Atta link. Ethnic minorities-man
2nd gen in Janukovitz
Kazakhstan
The Evrasian continent clockwise:
Drug related.
See them differently now. Mark them for later times.
The missing river: Saramanca
Yr aura scattered around (at will). 30m. radius easy.
We just can´t feel for you. Man invisible, remember.
A voodoo doll for petty gods.
Tourette syndrome
Cameroun
Smolensk
Minsk
Lichtenstein
N-Japan
Many of the Ancient Greeks in Byzans
We are smaller now: more quantities.

Eternity. Nations grouped and regrouped together.
Recreating historical standpoints.
Those who are not (as of yet) within you
won´t change the overall.
The hermit. (India).
All over Asia. Leaking military tech to Burma.
Someone old had to go.
Louis Alvares

Oct 10th 07

One of our men (London)
Died of old age.
From the abc to the xyz
An additional physical dimension
More than 300 nodes in the body
Corresponding points in the brain.
Color search. Sexual. Zero color left.
Heart 30+ colors found.
Muslims. Circumcision. Those who do that marked
And punished alike. Carve in the brain for greater satisfaction.
Few other sects. The same
How they carve their girls.
So that the woman in general can be led in a distinct manner.
Has to change too.
Enter individual elans from Japan, China, S-America.
Abdomen 300 millions run through.
Not to touch the inner organs.
China. Few thousand died. Most painful.
Failed to put a string in a point lowest in the gut.
Points central brain
Coated with light

Different sexual behaviour
Listed. More quantities.
Frontal lobe. Insert estrogen
(Two gases)
Color search, the brain
We use voltage from your very home.
Plugs. Up to 30v?
Giacomo Compostella
Linné
Root in Guatemala
    Neighbours
    Inks. (Honduras too)
    U.S.A. man
Cargo from Guatemala to Alaska.
Radiation as a tremendous magnetic force.
A lite on orbit.
The data it sends back
Images, sounds of Quasars, dwarfs: phenomena in deep space.
A reverse storm.
There were in a way two opposing forces
In the States. Remnants since before the civil war.
Protestants and Catholics,
Anglo-German stock and Latin American.
It just came hurling down on one of their men. (Al Pacino).
The war on Japan.
Use nearly the same frequency
We can narrow ours.
Intelligence. Offices in Mansjuria. Malaysia.
Singapore. Communications system down.
Their offices, one by one.
University-man.
We responded fiercely.

Point locus. Down.
Prison intact.
Building sites. In a cave below sea level.
They spread radars in the ocean.
Jet engines. Smoother design.
Used magnesium differently.
Their communities: Hawaii, S-America, N-America (Phong).
Insider in England
Some who didn´t help.
They were building a torpedo
which could accelerate at a given time.
Would approach below sea level
and accelerate when in thin air.
The sea.
Nacl compounds. Other
T++ V—
VVV
New batteries
+
(--)
Zink
(++)
-
Ghosts, magnified.
A sort of an acid bath. Dissolved.
History reveals itself.
Deeds, lies, deceit.
Detectors for under the sea activity.
A language system
Codes
Calculations
Drawings

Equations
Light wave intelligence
Waveguides
A device for systems network
Scaling up and down
A most dangerous knowledge. People imprisoned or
Under surveillance.
All too simple to reinforce the intelligence
And to create weapons if the rules are known.
A gas container. Over England.
A pressure balancer didn´t work.
Exploded. Were driving it away.

Oct 8th 07

Pakistan. Minutes after the attack on Mme Butto´s convoy
Faisal
Hussein
Al-Sadr
Musharaf
USA. Transition time over. As one nation for once.
Russia. Part transition time.
Old Europe.
Belgium
Netherlands
C-Europe
Spain part isolated. American Catholics :to leave old rivalry behind.
Kurt Vonnegut

Oct 20th 07

Few thousand Gopi-girls died within.
A.C Baktivedanta

The Hermit (NN)
We wanted to liberate them. (Self abuse).
Friendship built
Friendship lost
Girls of all times.
Instruments driven by electricity. Plugs in your home.
(Just turned on the TV) Circuit. Uneven flow.
Sudden voltage decrease.
Back to Euklid.
X, Y elastic
Engels broke the news in 1849 in his book ´On India´
That she was subjected to the Crown.
Algeria. Addicts of all times. Sexists.
Massacres during the Ottoman.
A branch of one of the pharaos
Arabia. Regrouped.
Have to evolve.
Polygamy, women´s right, circumcision,
Theocracy: the very structure.
Images: a pyramid in the making.
Villages in the Sahara desert, (N-V)
(will be an archaeological find)
Carthage standing
A mosaic floor under houses
. . . to bring images to life . . .
Images from within the library in Alexandria
Still searching. The fire broke loose
In more than one place, we care to think.

Oct 24th 07

An Arabic ghost died. 3d generation.

Lost some history, then.
Burma. Rangoon. Ghost killed.
Tech from Japan.
Cambodian borderline.
Laos. Iran. Response.
The hierarchy upwards
And from a point in time, down.
Wrong decisions-making early on
-so that those who need to, but don´t want to
Won´t be punished.
Highly dangerous.
So many of the old short of Essence.
Russia. A vessel in our (Icelandic)
Airspace. Launched at England.
Hit the wall.
Regrouped. Their Engels sorted out.
Isolated: lose autonomy for a while.
Kasparov given more responsibility.
Missouri.
A deed infecting a society
Corrected
Ionization table
Tried to repair the layer. (The gap, thanks to an experiment in the early 60ies).
Ionizing elements, gas, ight
Bound for a moment
Fell back.
It spins with the earth, but a bit slower. The gap
As it were, seasonal.
Millions of Russians through
Women marked
Addicts, homosexuals.

Engineers, doctors, scientists (see through them)
Thought they had found a loop. Through Spain to chavez, fx.
China. Hong Kong affairs.
Theirs only in theory for now.
Burma. Communic. System. Cldn´t trace it
Ours. Just codes
Their working class hero
Another.
Site
Japan. A few were calculating
Beginning experiments
Pre-emptives.
Methods of scattering
(Rebel heart)
Nearly every calculation of the last century wrong.
Doeppler, the speed of light (not as fast; the solar system
Thus smaller). Equations . . .
Few new ´beyond the pressure point´gases found.
Feed those full of anger and hatred with a mixture;
Glanular acid. Moss green for cannabis addicts (until there is no
Longing, no image in fact). Three in one or at will.
The same with other deeds Yr aura. Gathered in one place.
Eliminate movement.
Malavi
The better part of Afrika south of Sahara. HIV positives.
Have to
Touch them on ground. Marked: other pre-emptives.
Magnified. Grouped together. To control them better.

Oct 28th 07

Bloody blue

Russia used their women magnified. Enter.
Threw acid on the net.
Dirt man again
Cleaning the sours.
China
Managed to build a more powerful laser
Platform down
Vessels
Loss two(?) thousand
Massacre in the wake
Three men, every site the English knew of,
Peking ship, young and old.
Some images passed unnoticed within
Two Icelanders executed.
Aristocrats, Elite, nobles
An old woman in Sweden,
Eternity so entwined.
A young girl Icelandic. Killed someone on earth.
Edited. Experimented upon. Was supposed to be able to function
: Only to love and feel ashamed. Had to go.
Cambodia. Sites.
Heavy ions. New patterns.
My people. Either work within or face the consequences:
Prison or severe punishment.
Territories trans-ural, parts of China. The landscape.
Possible hiding places.
The English took some of their gear.
Can do it at home now.
W. H Boyer, Mathematician.
Aleister Crowley.
Lincoln
Bellman

Nobel
J. Cleese

Nov 11th 07

Mbdenum frequencies (Mo, 42)
Deuterium
Thorasin
Diabetes.
Tropical rain
Acidic.

Nov 19th 07

Speed	analog
Velocity	digital
X	x

Rb(Rubidin, 37) Rb + x=gas beyond pressure point.
Irridin 77+H+ lifeform
Mg12 magnesium
Mn25 mangan
Bill Gates
Lord Cromwell
Louis XIV (We, the people)
Charles Dickens
Bismarck
Hérge
Ign. Loyola
André Gide
Proust
Camus
Foucault
Léon. Actor

Alexander Dumas
Sarkozy
De Gaulle
John Nash
The specks hundred times smaller.
Categories for light
Masse
Charge
Density
Another astronaut died
Discoveries
Indochina
Old peninsula ghost died within.
Caused civil outbreak in Georgia, Pakistan, Burma.
B. Bhutto and the old elite.
Mohamed Atta. (Magnified). Perished, as so many that day.
Made for this sole purpose.-
Gogol
Tolstoy
Björnstene Björnson
Zimbabwe
Botswana
Ruanda
Nigeria
Padua
Alzir, Spain
Grenada
Nomad Transport
Somalia
Iran
Pakistan. Troops on the move
Red alert.

1/20 first generation dead in China
Ketamin-Essence. Find.
Russia
Man Finland
Siding w. the enemy. Launching site Russia
Slide show.
Moscow.
Burma.
(USA, India, part Bangladesh, north of Europe with)
Italy isolated. (for now)
Russia. New explosives.
On earth: exp. With small nuclear bombs.
Rearranging elans.
Pollution
Lake Malawi
Sea of Caspia
History
Abraham died in the days of old.
The good Samaritan killed.
Japan: the yellow peril
Found our frequency. Easy to make if you know how. Only ten components
Detectors in the ocean, dismantled one, exact copy.
Light by types. The same tech.
The defence wall build for old tech.
The security net didn´t hold.
They use
Fluorescent crystallizers differently
Can capture photons now.
For telecom. To trace, detect instruments in most of their men.
Their Hawaii man taken over. University-man out.
It submerged

A monk. Cambodia
Prisoners of war. Their catalogue of images emptied out.
Put into the search engine.
On earth. Industrial purposes.
British telecom service and allies.
BTS or nothing.
More flexible
More transparent
Japan. Lost their men.
Run from your elans.
The Muslims were impregnating Zimbabwe. Arabic trade.
Russia. Cannot see us anymore.
Turned on their old Communications network.
Incochine. A Base.
Melatonin and fermon also in the soul
Our astronomer
Was halfway through the galaxy
Minus degrees
Complained about his back before.
The helmet came off the suit.
Found out why. Yr vertebrates,
Scull; instrument. Insert. Spinal fluid.
Intron, extron, codon
An equipment
To take images from a large group at the same time
To prevent them from thinking in a distinct manner.
New treaty
The belt of chastity put on once in a while
Sex crimes.
Bernard Higgins.
H.G. Wells
G.B. Shaw

Luther
Calvin (died in the 30 years war)
Continents, countries done anew.
Morale and such.
Man Japan.
All drugged. Flower power mixture.
Individuals within. Just didn´t see real at all.
Just didn´t.
Hard to translate their images.
Based on hatred and fear.
No transparency
The caressing of a woman/man
Never saw them working
Pith and marrow. Erroneous.
Each and every one of us lost some feeling.
Historical turning-points.
We cannot (as of yet) run eternity through as a whole.
Have to canonize some points in history.
Erroneous for some countries. (-if only you had
Consulted us).
T.S. Eliot
Bill Clinton
King Arthur

Nov 14th 07

Photon-rich compound
Yields photons easily.
Scattering methods
Gas+light
Devices
Tony Blair. Needed a man in office

this fall. 10 years late.
The pelvis
Bones, breast tissues
Physical structure decomposed.
The States.
M-W
Someone who was big here.
Plotting against us
Few indiv. from a Ghost gone
The nation loses some
A grid.
Instrument over all of your head, inside all of your brain,
Down the spine
Develop new space suit
Midnight. Out of the atmosphere
In the shadow, - 45° S
Robotech. No joints.
Pakistan
Bhutto lost a group
They made her anew.
Musharaf.
Persuasion
Venezuela. Tech from Japan.
In a river-bed. It reflects sunlight. Harder for us to see.
Japan. Hirohito.
Equipment which detects radiation: activity.
Koturbinsky
Working around the clock.
-(we lost friends. I could be next)
Relentlessly
Driven by old hatred
towards England, resistance

towards any aggression in fact.
Don´t need to put up a show
to put him in office.
Tokyo news. They broadcasted what happened
on earth.
Intensifies our awareness
Makes us merciless.
Burkina fasa
Saratoga.
Moscow ship down.
One of their man.
Cabinet. Strong Pakistan relations.
Spain. Old imperial influence.
Russia. A peptid we hadn´t found.
Systems from us up with actual satellites.
Your natives fortified.
Witches brew.
Some crystal base.
Russia.
Within.
Head of the army
every institution:
every ghost; neighbours
transmitters along a railroad.
A site near Warsaw.
An uninhabited island
Tallin activity.
The Khan family: Mongolia;
Until she blossomed out of control
Belgium>Congo. Explains a lot. Somalia.
All the recipes.
The world is on.

They wanted to burn the atmosphere
and us with it.
Lactacid
On souls
Cholesterol added, He
Differently charged particles
tiny device
The right environment . . .
Pakistan isolated.
The rulers took advice from addicted individuals.
Within a ghost.
Maradona. Where football is religion.
V. Hugo
The Scandinavian currency (the krona) somewhat supervised.
Burma, Venezuela
Most of the sites down.
During the middle ages,
Buddha went astray. Corrections.
Con fu tse
Moral code revealed
-to give the woman her status
China: the coast-line on one side
And the inland on the other:
they have always had to bypass
a movement. Too many died.
Bad luck doesn´t come singlehandedly
A Hindu god died.
Softer for the body,
Which means it´s hard.
It just jumps over.
The book of maidens

Nov 18th 07

Chile. Activity.
Syria. Soldier-shields. Advanced.
Kalahari desert
Cuba. Guns
France. Experimenting.
Near Paris. Inside the underground system
Near the Belgian borders. Turbines.
Just watched them.
Sites down
Persuasion
Part De Gaulle, Foucault.
The other side of richness.
Autocracy.
Near Beirut
South Libya
Egypt
Russia, (a few cut themselves off). Moldavia
Austria (3 places)
Denmark.
-pre-emptives.
Everyone wants to try out this new technique.
Perivarod
Some have weapons programme
Others are allowed to- under surveillance.
Geometric values
Smaller eq. (due to more compressed gas,)
Faster
Capacity
Samoa islands
Japan. Samurai link

Yucatan. Root Marley
Detroit; neutronicity.
Pay the prize
Collective China
Opium. Dutch within. Dismantled.
S-America.
Isolated.
Windows closing.
The fleet
An aftermath really
Day of the lord.
Sudan
  Omar Al Djahire
Somalia
Djibouti
Philippines
Tech destined to Maroc
Kosovo, Montevideo.
Looting.
Moldavia
Kaukasus
Moving eastwards.
A fleet of a thousand.
Point space from the altitudes
The fleet scattered.
A site. A tech we hadn´t mastered.
Ionic force.
New energy source.
Shuttles much faster.
How to tame it.
C-Evrasia.
Between the two great lakes

scanning the area, securing routes.
As if a wall moving:
those who get in the way perish
Montesqieu
Tesla
Nitroglycerine + compounds from us
Greater explosion
Any resistance fought vehemently
USA on the werge of dicovering this new tech.
Experiment in the void.
Manned and unmanned spacecraft
exploring the galaxy.
Tracked a signal from a distance
Probably sth. We sent up earlier.
Kleine Götter

Nov 20th 07

(BBC):
´Kazakhstan
Not so much production
but services and consumption.
Microloans. Finance banks.
The capital as a model centre of the region
-no taxes´.
Cyclones, hurricanes.
We influence the weather tremendously
Russia. Run through the directory.
The rulers. We hurt them until they
Do the right thing. Not for war.
Many of the emperors of Rome,
Some of the kings of Europe,

Some of the popes
Highly selective families at first.
The fifth evangelist: Bach
Beethoven
Mozart
Kopernikus
Kepler
Galileo
H. G. Wells
Ch. Dickens.
Dali
Max Planck
Riemann
Rembrandt
E. Munk
Goya
Some dukes and doges
All too many presidents, ministers,
Heads of important institutions, organisations.
Baath party. So corrupt.
Old dispute. Hatred
M. Atta used a flexible blade.
Individual within Atta also within S. Hussein.
Kosovo. Serbia
A safe haven for many Nazis.
Lusitania.
A month ago
Few women from Colombia
Were granted asylum in Iceland.
One of them was with the opposition,
Watching the activity, experiments above.
A young man from C-Asia, who had been working

Within for the Asia population here. Went to the high-ups
In his homeland. (Surinam).
Jack the Ripper. Found
Not a practicing doctor
Don´t know how he passed unnoticed.
So many men died suddenly
During the plaque (1340ies)
Never had any correction whatsoever.
Sir Walter Scott
Charlemagne
W. Whitman
D. Alembert
W. Blake
Feudal lords of all times
Great many Rabbi´s
Dagestan. Confrontation
S-Arabia
Some who participated in the blue mosque incident.
W. Reich, of a pioneer nature.
(from the book ´cosmic superimposition´):
Orgonomy (p. 4)
´Facts and interrelations in superabundance
Bions
Biopathy rests on the decomposition
Of the living organism.
Orgone accumulator, (156).
Sufficient charge lumination.
Self-regulatory children´.

Nov, 22nd 07

Putin: the western powers are playing

Dirty tricks behind a weakened Russia.
Pkistan kicked out of the Commonwealth.
It´s just old tech. Heavy and slow.
Not nearly as powerful
A ship. Vessels.
Karbala
Ohio. Cleveland
The states. A drug of fearless
Arabia. Found 3d force negative. (From Vietnam?).
A fearful weapon.
A ghost out. Were closing in on him.
The man found. Losses.
It fluctuates, repeats itself geometrically.
Softer. (not as dense; yellow-white)
The denser light just goes through.
Can detect (see the pearly users now.
Didn´t know how to search).
Over England..
From Arabia.
They just figured out how. (a coordinate system).
Someone just seized control.
A genuine massacre followed.
Were working behind our back south of Sahara.
DRC activity.
>civilians.
Damascus.
To teach you, (the Icelanders) how to take care of your domestic affairs through the body of Christ.
The pains of learning.
A thousand recruited.
This war is just killing us all.
Activity in the States. Neutronicity.

Off shore of Manhattan.
To the last man standing.
(-where have I heard that before?).
We have transgressed it all.
Old high ups in Europe have put an effort
into bringing this war to an end.
A transitional (metal) detector didn´t work.
Negligence. Carelessness.
We are becoming short of heroes.
Oppositionsgeist (to use terminology from F. Lange).
When something happens in the heavens
People on earth is strangely at ease.
By command.
New Elans.
Dakar
Idi Amin
French
Neighbours.

Nov 25th 07

Nebraska. Activity.
Individuals, magnified plotting against,
Leaking info to the adversary.
Dodi Al-fayed.
2 years ago.
Quite a few adversaries.
16th c. magnified France.
(within Napoleon, half of the western hemisphere.)
Traced to Mussolini.
Measles
Japan. In Melanesia.

Russia. Lite up. Attack on a building site.
Fierce response. Magnified by numbers.
Ever so many have been magnified throughout history.
People of importance, people in the spotlight. Loved ones.
Up to a thousand within. Lose part self.
New elans. To give those who lost something back.
Enter list of names. Perish. Not one by one. From within.
Decave

b flat

F sharp

More precise

Nov 26th 07

A kind of an ambassador. English. Within Kasparov.
Killed. Fierce response. Traced around the world again.
Treacherous, plotting against us.
Collective Russia, as if categorized within, (by us).
New elans. Egypt Took a chance.
Opened up one of their oldest. 3000+
Tribes by the times of the first pyramids.
Our Icelanders in Russia. Killed as spies. (20?).
The Russian embassy in Rvík.
Discretely observing.
Pay in like manner.
Our Eastern Europe –Asia population.
Mullah Omar. Someone within him
Was with Muhammad Atta.
We run the show there now.
Instrument to hanter with the soul.
Confiscated.
Cairo

Ankara
Grand Canyon
Biscaya
Med Rep
Some cuts. ′gloves′. More precision.
L and R hemisph. Separated. Outside of.
Points in each of the hemispheres. If cut through,
the body loses aureal receptors: one of the peculiarities of
your nation.
Elementary beings. Primitives. Nightly visits. Unattended.
(Our young knew when the elder were busy elsewhere.
Their negligence).
Craftsmanship supposed to be in the hands of a chosen few.
Nature spirits. With a decadent flavor.
Youth demonized. Spooks
As self-appointed emissaries haunting the living.
Attacking them at night, distracting them in their daily affairs,
following them off rooftops, over board.
The world. How we touch the brain.
Old habits, practices. Movement eliminated.
Frontal lobe, midbrain
Tiny movement in points actual. Light run through.
A sphere of our own.
Energy fluke
New kind of light.
It sort of jumps over.
It recreates itself in another point space
at a given interval.
Charge
Yields or attracts particles.
Fast-slow,
at a distance.

New hardware.
Softer on your body.
Faster vessels
More efficient. (industrial purposes).
Network. (sound and sight). Programs so than none escape.
Bulbs
Transmitters
Receivers
Lenses
Detectors
Core yealds particles>lifetime
A machine. Over areas at great speed.
To secure peace.
Add a flavor.
Add: greater satisfaction.
Add: More intimacy.

Nov 28th 07

Beyond recognition.
New Elans.
From the mountain. (Lebanon). Addicts throughout.
A garden of Sativa on an estuary.
Traced to Kazakhstan, mnt Altas, France.
Raid
Panic in the Arab world.
Romania
Kirgizstan
Besancon
Rome
Copenhagen
On English soil.

The States. When inactive. Impossible to detect.
Someone on the line.
Insider. The inner circle.
Treason.
New green light.
Near Damascus.
A town. Ruins. 3-400.000.
Mesoamerica.
Haiti. Cuba. Raoul. Miami. Bolivia.
Laser tech has evolved: twice the distance. (other color combination).
Light divider
Green laser.
Over a large group of people.
Medical purposes. Industrial.
The Barrel. Void. Greater speed.
Decrease resistance
25 30

2
12 0
Lost an element of surprise.
Near Dardanelles.
Cincinnati.
Someone too big.
Was practically everywhere.
Early 19th c.
Within Pavarotti,
Elvis, in Russia. Had worked here.
From Charles Dickens onward.
Points. Nerve-ends.
Clustered,

Multiplies capacity.
To bind them from within.
To hurt them for their deeds
(to make them stop).
Interrogation technic.
Easier to supervise
Intertestamental times.
Essenes. A part in Rome, Jews.
>Zohar
Scramble their won intelligence systems.
People made to feel in a distinct manner
Image control.
S-C Africa.
Australia.
Millions through.
One of the highjackers had a knife.
The Americans didn´t tell us.
Your teeth. (The Gum/Palate). In six parts.
Russia. Muslim protector.
A Frenchman.
Two English. Old. Kandahar.
The war on America has ended..
Your aura. (Specks, parts of it), when within.
Used in the workshop. To find deeds, to move light about.
For industrial purposes.
Foucault. L´antiquité, une profonde error.
Descartes. Celebris promissor.

Des 7th 07

Keramics. Keramic essence.
Elasticity.

Leitmotive
Russia. Their communic. System.
Use Mercury differently.
Some of their lamps for industrial purposes unpresented.
Elaborating ours.
Conspiracy
East of Ural.
Killing us from themselves.
Fierce response.
New material
It just weaves itself.
Under, over, from top to the right
Women, young of the world.
Addicts. To make them quit.
Old Castle principates in England.
Corrupt.
It looks as if it were alive.
Circulates at an interval.
Girates.
Scales.
Influence of Judaism. Coptic ritual.
Partly in early Catholicism,
In pre-Mohammedan era.
Wiesbaden
Nassau.
-Las Vegas.
Wichita.
Mujahadins.
From Lebanon, Sardinia.
Wind in the sails for Hisbollah.
Rebels, resistance, uprising.
Russia. So much for the transparency.

Just made themselves new ones.

Des 9th 07

A million women through.
Few thousand died. From Macedonia.
Badly marked. Weren't magenta.
Can´t have any other nation work here.
Observe too well our activities, tech innovations.
To evolve as a second pole
Helions
Mercury ions
Mangan ions
The lava is pouring out few hundred miles off the coast of Sumatra.
1000m deep. Coagulates. Therefore earthquakes one in a while.
Rain. Snow. Sour. Radioactive in fact.
The USA put up a show.
Part Elvis. Part J Cash. Other.
A ghost in the machine.
No spare parts.
Losing precious time.
You´re within the plasma.
It vibrates. Fluctuates.
Anchored in the physical
Energy cube.
Higher voltage. Fast measurement. 300.000
(40/100 pr/sec. Charge, nerve-ends, physical quality).
It doesn´t take long to run a million through with this tech.
The losses are reversely tremendous. No way to stop.
Miscalculations, failures, something not foreseen,
Polarity of power

Etheric radiance
1:20 dead to the world

Des 10th 07

Myriad magnets.
So powerful.
New hardware. Better resolution.
Turn nasty. Emotions change. More focused.
(From an English newspaper: sliced cactuses for ´hangover´)
We were rearranging eternity through historical standpoint.
Christians from Christ, firstly, by Bacon secondly.
During the process in China, (a standpoint arbitrarily
Chosen by us): a whole generation died.
It is just impossible to hold the hierarchy undivided.
Man is the limit.
Better part of the Ming dynasty just wiped out of history.
13th (Xia dyn)
Until the time the Mongols where flourishing out of control.
16th c.
We´re changing history fast.
Layers
Specs
Vapour
Like a crocket ribbon (old women know what that is).
In one eye out the other: from the inside; perpetrators found.
They developed a fire wall. Watch them fry.

Des 12th 07

S-Ukraine
Panama
Suez. Under our Control.

They won´t attack our fleet.
Aerial weaponry.
Chemical agent which ´melts metal´;
Decomposes distinct materials.
New force. Attracts and repels.
Your feet. Light of movement. Split up. (strings)
Engines on containers, equipments, probes. Just driven around By remote.
Africa. Premature death.
Because of too fast metabolism

Des 13th 07

Battle of the houses.
A capsule close to heart. Elan from diseased, gas;
thus the quantity.
Could extract specs from your collective w. the right mixture.
Put in every house. Few hours later: Jordan: killed you. (your specs: our see)
And others. An English dearly beloved.
Addicts, rebels. Fierce response.
Algerie. The same. One of your Vestmanna-islands girl killed.
Tibet
Cambodia.
Pre-emptives. Persuasion.
Tele-lies. Deception and like.
Zimbabwe. Half of the nation addicted.
Soft drink. Compressed leaves, opium mixture.
Russia. Gave in, it seems: man Kandahar.
Armed and addicted.
Argentina. Wome within Maradona.
Cocaine. A warehouse. Had become the biggest

Exp. Country in S-Am.
Iran gave in. Abandoned their nuclear programme.
A bio-Robot. An individual soul, (guilty of a most severe crime),
sorted out. Made to be within you in full size.
His hatred, his anger, his feelings drained.
(Neutralised). Memory out. Sight hantered with.
Some of us have to see for him.
Programmed.
The individual therefore out of eternity. Lifetime.
Post-war trauma.
History synchronized. Have to fill in the gap in China.
3 individuals from Iceland executed
Brought here from to England.
An Icelandic woman seems to have spoken outwardly
to Swedish colleges. (Leaking info).
Negligence from someone working within.
A woman who spoke to a man from Belorussia
eho followed few Chernobyl children. (He fooled her
Apparently. Passed significant information on).
Attack on the S-Pacific fleet. Burma. One house.
Den found. Response.
Some of us within individuals in the States killed.
(Supposed spies).
Here. Two from the American embassy opened up.
Activity around the Bermudas. Try out the effects of the sea.
Activity off the coast of Venezuela.
English and Canadian civilians in the States. Killed.
. . . there was a covenant . . .
Bio-experiment went part wrong.
Few hundred handicapped.
Two from B Higgins had to go.

A gadget
Lighten up few square kilometres.

Des 16th 07

Someone within Elvis who worked here until Visco.
Plotting against us.
They have been onto us for a while.
Microsoft. A card up their sleeve.
Aircraft. Two in crew. A pilot and a navigator.
Missile tech. Long range.
Moldigliani
John Dunne
Mohammed isolated.
China. Someone seized power. Sites.
Ninive.
A city-state in Mesopotamia.
The States. Locked and loaded.
Diplomacy to the extremes.
Civilians dying.
Beckham. They killed some from him.
So many individuals from us are within over there.
Their fleet. Liberia , Iraq, S-Arabia,
S-Pacific, S-Korea, A base in Brazil,
In the Florida Swamps. Activity.
The experiments on Yucatan. Their responsibility.
Persuasion. Their men isolated from within.
New elans. Magnified.
Man Kandahar.
Two from China.
Old Byzans within. 2/3 addicts.
Uruguay. Thousands.

Pan Ama. Sites.
Accident in outer space.
A Shuttle just exploded. Unknown cause.
The nations of the world are uniting
against the commonwealth.
Synchronize their actions.
Dialog betw. America and Japan,
America and China.
The cruel dictator.
The worst of structures.
Communications systems within satellites on orbit.
-(We are dead in the soul).
England. Opposition. Pacifists.
Fear. Disillusion.
A group of radicals had to go.
Ill fated places.
Districts here
Streets there.

Des 20th 07

Fluid around mouth. In Genitals.
For more thorough search.
Youth in Tibet practicing homosexuality.
The finest tissues. Throat. Lungs. Capillary.
Lowest part of the cortex.
To ´write with light´. Industrial purposes. Other reasons.
For smokers. To hurt them to make them stop.
Run through. Thousands every second.
Machine failure. Few thousand died.
Your house. Your brain. The greatest cavity.
Here. (Reykjavík). Someone within the Finnish consul.

Gazing up the skies. Leaking info.
A French woman. The same.
The English didn´t choose this out-of-the-route island
for nothing. Experimenting around the clock.
The Ottoman empire.
The Eve of industrial revolution.
Mostly from Indonesia, India.
A great old family from the latter.
Princes and princesses. Some hurt
themselves beyond in the throat.
Had to help them over it.
Americans started suddenly to kill us from themselves.
Few great men. People of all ranks.
Lost a man from the war cabinet.
Launched an attack on some of our bases in India.
Fierce response.
120 sites and bases in the States attacked,
killed them mercilessly.
A vessel over your head, (protection).
Shot on English soil. The shields over England took the blow.
Equipments and machines confiscated.
Russia tried in vain. Germany, Austria.
Our loss thousands.
The money changing pockets. Traced.
Our tech + some tech from Japan.
They were developing wireless.
Underground desert activity, remote hiding places.
Drawings and calculations.
Fatal accidents in the wake.
From the inventive site.
More elastic.
Light disappears and reappears at a distance.

Materials heated (few hundred degrees). Fundamentals.
Your body senses the heat. (to say the least).
Mostly magnified.
You hold 7% of the magnified in the world.
In England
Old and high up. Traitors.
Someone else backing him up.
Most of our men lost some.
We gave the Americans to be within our men/us.
They did the same. To strengthen the bonds.
Many of our magnified.
The young lost friends.
The safety net (England) came tumbling.
Didn´t know how the predecessor (now dead) thought.
The States. Vessels in Sahara desert.
Allying with their former fiends.
Hereditary tradition
When was the point of no return, really?
A portable equipment connected to the computer.
To wash out memories from an image to an image.
(We ´know´ it is wrong or right, but not why).
Jesus within Muhammad. (8-12th ? generation).
Different costume. Different habits.

Des 22nd 07

Someone within Bush. Organized the lot. Took the flight.
New elans. 100 magnified. From the States.
From Argentina. Tons of cocaine. Been onto it for a while.
A Cannabis farm from the early 70ies.
Traced to France. England.
Few magnified gone.

Cannabis Sativa was planted on American soil with
the first Spanish settlers.
Cambodia. Sites down.
Indonesia. Killing us from themselves.
Queen Elizabeth lost one more.
Enter man Indonesia.
Conflict throughout the ages.
Indonesia.
Attacked Borneo base
Another in one of the islands. From all directions.
Were preparing take off.
More than a thousand dead.
Response.
The population on an island wiped out.
Places here and there.
Traced to Thailand, Cambodia.
Old Mongol chef, A Buddha, Hindu branch, Sultans.
A third of their man who entered earlier this day gone.
The Ganeshi (Elephant God) family lost a branch.
The Dutch (the white buffalo, as they were called), sympathizing with their former cornerstone of their empire.
Scanning the islands
Trying to run away. Eaten up by the flames.
Sumatra got hit badly. Waste areas.

Des 23 07

The confrontation 2? Days ago with the States just an exercise.
New radar tech, communications tech. Activity all over Mexico.
On a Ranch, by the seaside, in the forest . . .
Attack.

We hurt their men to call it off.
Counterattack. Radar down.
New Elans. Mexico.
Matt Dillon.

Des 24th 07

A group in this conspiracy called on some Arabs to attack when the whistle came. Few Vessels came in our back
in the Middle-East.
American vessels in Bagdad took off for Basra
but were called off.
Addicted demagog. Trading Heroin. Military transport..
Our soldiers. On a hostile ground. Easy to be misled.
Spain didn´t move. Russia neither.
Our timing perhaps.
New elans Brazil.
I. Ayola
Aztec. A Farao within. (From a city-state in N-Africa).
Tracking death pattern.
As if by a numerological precision.
Wrong/bad decisions-making sometime in the past.
Tolstoy. (Some of his magnified). Not killed by the Soviet.
By the States. One of these devious political moves.
New elans Holland.

Des 26th 07

Ions ahead.
Ionization method.
Equipment. Image. Those who know.
So many innocent die within.
The image denotes or signifies sth else too. Inaccuracy.

Massacre of the great. From a country to a country.
Fascism and Nationalism seem to follow suit.
Doctors (hantering with the soul; do not want them to kill us from within).
Scientists, engineers. Enemies at large.
Military on the move.
Russian torpedoes, (on ground).
Changed direction.
Paralyzed.
Pyrrhic victory or
Measurements beyond.
We were losing too many good men
Because of progress in the medical sciences.
Feared they would develop a mighty weapon.
Every battle won these last months
demoralized us to say the least.
Ion shield
No light penetrates.
Jasmine. Age-old Arabic princess.
Addicted from the foremost generation onwards.
Practically every spirit has lost some great these last months.
Mozart. (The one within who composed the better part of the Zauberflute). In the mouth of the Russians.
Adm. Nelson, W. Blake, Churchill, Lost a branch in the States.
S. Becket. France.
An endless list.
Of wanting to fail.
Erroneous thought.
The States. Anonymous. Preparing attack on ground.
Liner crossing the Atlantic.
The intelligentzia. Army ranks. Few magnified.
Can´t get close to them as of now.

Top officials. Killed men of importance from within. (Thousands).
Until they surrendered.
Gave us the equipment we wanted.
They cannot kill us from within their own.
Russia. Their army generals, officers. Under surveillance.
New places of military importance.
Someone within Newton a week ago: it will be our death if we go against the States.
New elans. Magnified.
Rings (force fields) around some quadrivial places.
France. Some pre-emptives.
Warlords. Those with the knowhow.
Your blood vessels erroneously done. Corrected.

Des 28th 07

Two months? Ago. The Russians shot down our polar machine.
(It tracked down comm. system network. Other.
Works with earth´s
Magnetic force. A craft. Over areas.
Burma. Comm. sys out.
China. Japan the same.
Russia. First battle lost.
Have evolved. We´ve got nothing like it.
East of Ural.
The grandfathers.
Lipids
Lost an infantry.
Bulgaria.
Syria-Gaza.
The Americans refused to help Israelis

The bravest of the lot
They´re tearing us apart.
Too many posts.
Fewer and fewer believe we can hold it.
We feel for our people in our men.
All terrified.
Lipid experiments within.
Tau. Sigma.
It sort of ´denies itself´. Refuses to attract.
Thailand. Ambush.
Their man found. Central Evrasia. Muslim.
Experiment near Aral.
See within their house.
Pre-emptives.
Here. Two from the Russian embassy. Spying. Dead.
A fellow countryman. Within an Icelander in the States.
Dead. Was working within.
Cannot scan areas as before.
Acid explosion at that time.
It just burns the atmosphere
in a radius. (and us with it)
USA. Their African-American. Activity within.
American-Chinese
their American footballer.
A syndicate.
Still activity in France. (S-coast)
Fr. Mauriac lost some
C. Rhodes
Wpns confisc. They sabotaged it before.
Experim. in a cave. In old mines.
In some underground atomic-bomb shelter
FIFA arranging matches

Argentina. Drugs. Transp. in a container to the east coast.
Costa Ricans distrib. for the whole-sale. Bikers.
someone in Bush. We want them all.
The Great Greeks went Byzans
B.Bhutto. She was the reconciler.
Bhutto dynasty. Nepal royalty within.
Consuming, absorbing
Some of their drawings, calculations.
The system within (and around) too slow.
Took too much time to verify.
M. Foucault. Someone died from him.

Des 30th 07

Representatives from 50? countries.
Have to split up this assembly.
Detroit
Kremlin fortifying
Arabs waking up.
Japan
Losing battles.
Accumulators.
USA managed to fortify their air defence system.
Lost our ship some time back.
One of a kind. Coordinate blast from the altitudes.
Some called it the mothership.
The platforms are out of date.
the ´eyes´ useless in some regions.
India on its heels.
confrontations along the borders.
Our loss might just set them free.
We fucked up eternity.

Hold most of the men.
They may not know anymore:
those who have not been touched by you.
China. (A woman in the states) Stealing tech from USA.
Someone within Bill Clinton leaking info to Europe.
Battleship Airborne.
Her first mission.
Russia.
Their spiders,(new tech defence system).
Sites. Radars and such.
Securing airspace.
New elans. Their new man. (Made to consume
their war cabinet, scientists, men of importance).
Murmansk. their top building site there.
See their army.
Their archies aren´t strong enough.
just can´t reach the altitudes.

Des 31st 07

Curt Cobain +
His countrymen killed him.
Suspected of being a messenger. (ours).
USA. Their man found. In a forest cabin. Boobies all over. From above.
One of our Ghosts got caught. Dead.
Their man out.
He was practically everywhere.
In the most important houses.
Lost his Russia branch.
His individuals try to gather what he knew.
We found out how to ´open up´ history.

Ehen will they?
The last days for so many of us
They make us (Icelanders) forget what we have done.
We were shaking like leaves.
Just blanks where images of preference were.
Accordion
19th c. ghost USA died from within.
Shot their highway down to gain time.
Automatic search and shoot.
A flair
Catalyser
They know what we can do.
Throw toxic on ground which eats up vast areas.
Their search more elaborate
Their archibald reaches the altitudes.
Milwauki
Montenegro. Serbia. Kosovo.
Wrong decisions.
Slovenia

January 1st 08

War on America.
1:5 of our army
Airborne down. Quite a few.
Their air-raid gun confisc.
Their loss tremendous. Hundreds of thousands civilians.
Some within Obama, L.Armstrong. Man Bronx out.
Their infrastructure doesn´t work for now.
Released their prisoners (us)
Polinesia. Hawai.
our S-pacific army. Australia..

Either with or against.

January 2nd 08

Russia.
The men we have here
are handing down their power.
New men in charge.
Their defense system unprecented.
Impregnating Turkie, thus engulfing the Sea of Kaspia.
Byzans remnants.
Started to kill some of their own. (Who were within).
Securing a route from the borders to one of their man.
New elans. Their man.
Endowed w. many magnified. Maps, plans, webs.
See another man now. (northwest of Mongolia)
another man in the making.
Use the centrifugal force
to throw their webs higher,
for a greater spin.
Shorter wavelength.
Hand over some of their weapons.
They sent sth for comm. netw. sys. on orbit
with their satellite network.
we might need to send manned aircraft on orbit.
Mexico isolated.
Tatars. Gypsies.
Jews. Italian, Austrian, Russian stock in the States.
Other minorities.
Magnified: Politicians, partymembers, high ranked military officers,
scientists, artists, media people. People of social importance.

Increased your capacity.
cavity inside the scull.
Tubes.
Burma. Combat. securing areas.
New Vessels. Bigger. More possibilities. Throw webs downwards.
Some gone from Clinton. F. Roosevelt.
To take over their institutions, ministries, committies.
(In a solid analytic language):
from the generatrix to the conic.
Ellipsoid: Red.
Ellipse to a point or from a point.
Projecting light from a plane, to a plane
or like ripples on a water.
Throwing ripples of light in all directions
or bulbs of light on target.
Geometrical forms as moments, repeating.
Like helici gone mad.
the densest light. Deadly.
Inverse, everse. Depending on a function.
A spiral unwinding from or winding onto a circle. (involute).
Curves of light peculiarities of shape
A spiral sweeping out circuits. rolling cycloids.

Jan 3 08

The States
Bought some tiny instruments
from Japan for their defence system.
Individuals from most of their men, magnified executed.
Old Europe. Germany. Old National Socialists.
China set an example. Gave us hints about

addicted individuals they knew of in Indonesia.
A tenth of Russia still holding back.
Old Central powers figures, to use WWI terminology;
Germany, Austria, Hungary, Rumania.
Russia. Experimenting with heavy ions.

Jan 5th 08

A device. over areas. Charge. Marked individuals.
Hell on earth.
Russia. Their loss tremendous.
Traced to H.Chaves,
Syria, Bangladesh,
Yr teeth (roots)
yr skin (ins. peraffin)
for prisoners intact.
Out through your tongue, inside through face.
for great quantities.
To play one image against another
They know they can´t beat us in the pearly.
Preparing an attack on ground.
One of our great. (in Al Pacino).
Our ´insider´ leaked everything to them.
A group in England, links all over Europe.
(Brought the technique to them)
Boswell

Jan 8th 08

A controlling device over every major city
on this planet.
Conspiracy. England. Political opposition.
Man magnified within Cromwell.

Seized power during the war of the Roses.
Dictates the lot.
Corrupt board
Few of their great.
Hiding evidence.
In Andreas,
in M. Thatcher.
Dagger dictator
Chief of police

Jan 10th 08

Obama . . . Some within Bush. One of his great.
Few hundred +
Distinguish between big and little Russia
Her ethnic varieties, and the Kremlins, resp.
Youth from wherever.

Jan 11th 08

SW China
Tech from Burma.
Specialized in Booby traps.
Air-space unsecure.
When inactive, hardly traceable.
When activated; they open up as giant claws.
Started killing those of us who were within.
We dropped a bomb.
When it hit the ground:
like an ever expanding circle,
the atmosphere burning within.
(Child with a schoolbag, running)
Wpns factory found. Out.

We experimented on some of your children.
3-4 years old. To make them not want to do sth, and reversely.
Implanted them with base feelings, iyw.
Few hours later one of the children died.
Most painful. Didn´t get vital nutrition.
Some of your girls.
Hormone excretion prevented.
Femininicity.
Small cuts in your genitals.
India. Attacked. Rammed the shields.
All reasoning is gone.
Brazil. Activity.
N-W of Sao Paolo. Pre-emptives.
New elans.
Calculate differently.
Intelligencies
organizations
corporations
: that they be under our hat.
The conspiracy counted 21 nations
USA, Russia, Turkie, DRC, Arabic nations,
We´re building a new craft for space travel
(Last generation just exploded. Unknown reasons).

Jan 13th 08

Those of us whose teeth are missing.
brought to you.
Car companies. Fixing prize.
Galleries. Auctions. Know beforehand how much
the buyer is willing to pay.
Diomedes

Jan 14th 08

Yr. system turned off.
Prisoners intact,
Adjustment,
people bound within,
on again.
France. Used them when yr system was off.
N. Sarcozy. His Arabic connection. Lost their mask.
The Arabs killed them. Lost one spirit. Magnified.
Black diamonds
Glow from within
Talibans. From a village to a village.
Addicts marked.
Our base Suez. Attacked.
Age-old Arabia opened up.
A family from the peninsula died.
Sadducheans.
Will be regrouped with Pharao family.
Four days ago. New ship.
Works wonders.
Force-field.
To secure airspace. (To pick them up).
Cambodia
Burma
Japan
Experimenting. Somewhat unprecedented.
Most of the houses known.
Drawings. Calculations.
Burma. New elan.
Lost him some time ago.
Lao Tse

Deep rooted hatred towards us.
(The India subcontinent)
Your jaws. not functioning
as they should for us.
Luxor bombings. One of the organizers
was within in one of the Twin Towers highjackers.
Japan. Osaka.
Under sea-level.
Site. It sort of galvanizes.
Rocket technique.
They destroyed their site, rather than
have us loot it.
Curious Cl - compounds.
Organizers, scientists and
technicians responsible executed.
Few generations up.
Their market capitalism.
Economy. More transparency.
The armed race
Here is my will my testament
brought to you and herewith sent
out of the wilbury, lonebury
my cries out in the desert
North of the Himalayas. Attack.
The ship down.
Loss. A Ghost. Boswell.
A division. Thousands.
Indian subcontinent.
India. Had us from the start.
Killed Ghosts and magnified from them.
Nepal. Neighbours. Old Khan empire.
Türkmenistan. Russia in fact. Ignited a bomb

as the fleet moved over.
Desolate area.
Bangladesh. Bengal. Bhutan. Sikkim.
China. Chausescou within one of their men.
Ceramic. The Russians found out how.
Started suddenly to kill us from them.
Their man found. Out.
Old Europe, as well as the States. Lost some.
Dialogue with the U.S.
Polar cap.
Excited by degrees
Arabia. (Muslims). Their spiral spin
was to the left. Conjoined to the rest of the world.

Jan 22 08

New ship
The Avenger.
First combat.
Minimum casualties.
The bombmaker found.
Forest Whitaker
Armenia. Man within
Our spinster

Jan 26th 08

USA. Their resilience stimulus.
Activity. Sites in New-Foundland,
Greenland. The Danes didn´t tell us.
Russia.
One of our old and great dead.
Got trapped.

Closed on him. A foxtrap.
Cameras in our backjard.
Machine parts.
All too easy to smuggle them.
Via Sunderland.
Bobby Fisher. Magnified. Our forgetfulness.
Spied on us from the start.
Days on end.
Some great within.

Jan 27th 08

One of the Ghosts within Musharaf dismantled.
How Muslims cut off their Christian, hindu,
jewish, buddhist heritage.
Fatal incidence in the wake.
The Monk Burmanese dismantled.
Their army lame for a while.
New elans.
Two Ghosts hold their army.
Russia link.
Snipers.
Their chemical agent found.
The ambush last fall.
Partly addictded demagoges.
The Chinese were strong there
at a point in time.
The Mongols. (Since their invasion).
Still present. Just changed costume.
Suharto. Magnified. Enter.
Root Elvis within.
Neighbours.

History recreating.
Lesbians. Harder to find.
Found an idea. Your mouth.
Clitoris recreated. How they suck on it.
Lose sexual potency. Their place in the sky.
Here. Kitten problems. Yr. youth. Yr. girls.
All of them nightstalking with some teenage boys.
The physical accordingly
a bad boy toy,
or a bad girl pet.
Sucking tarts.
Child abusers.
All too many of your countrymen
have injured the physical.
Memory loss, movement, potency,
intelligence; brain damage.
Various reasons.
They might have felt that the victim,
(the very physical), were somewhat
responsible for his or her death.
Just didn´t like them.
Not the right clan.
For the fun of it.
To get away with it.
And the elder, who acted
out of ´necessity´, as if by ´command´.
The individual approach, so to speak.
Elementary beings, as was said.
The last of the nations to be embraced
by the eternal dimension.
4000 of us had a lifetime.
The nations of the world will have to

upheave their individuals
in a distinct manner.
Persuasion.
Trials. Long and short. Muslims.
S-America. Venezuela.
One of their man blasted.
Couldn´t risk our people.
A month ago. Karadzik. Blasted.
An empty house for a while.
The Americans were infiltrating him.
Experimenting with flammable gases.
Taming the energy from radiation.
Abraham. Assasinated in OT times.
Baal. Therefore the endless debate
about ´the binding´.
The Koran. 14.45 Abraham:
swift is the reckoning.
´You lived in the dwellings of those
who wronged their souls´.
´To stare in consternation
with hearts utterly vacant´.
Vesuvius.
Poisonous gas. Many of those who were buried in such a strange ground are within.
Your aura. We break it.

Jan 29th 08

USA. As before. Started killing us from within.
Few great dead. A Doctor. Was within you
from the start. Lost some knowledge henchwith.
Response.

Obama. His heart practically vacant.
Lost those who were active within.
Arms producer found.
Their men with the know-how.
Bermuda. NY. Chicago.
Securing air-space.
They have a superlense now.
Man Cameroun. Lost a branch.
Was in B. Obama.
Mr. Bush. A day ago or two: ′We have to fight the tyranny on Earth′.
Their Native Indian, their Sino-American.
their African-American. (Actually in prison now. East coast).
Throwing instruments, weapons up from them
One of their Ghosts dead.
J.E. Hoover.
Watch them gather in one of their man.
Many of their great have had to go.
Some individuals who were within their great.
Made them lose in spirit.
Persuasion.
Handed over many of their equipments.
We′re experimenting in the countryside
South-East of Reykjavík.
All too secretive.
Gelfand
Türkmenbashi
The Edge
Confronting nations have to lose in spirit.
Until they start to cooperate.
People brought to you on a string.
(Your aureole for now).

Grouped together by a country, ideology,deeds, or at will.
Increases capacity, efficience.
Too much pressure. Some died.
Ghost from old Jugoslavia, others.
For age-old history:
volume of charge
heat, gases, biochemicals
Russia. Up in flames.
Some of their resources.

Jan 31st 08 Morning.

China. A coup few months back.
Mandarines losing. Another dynasty.
A prediction some time ago.
Distribution of arms among civilians.
At close range. Lost few good men.
One of our great. (Was within King Edward).
Nuclear blast.
Found one of their men. Brought here.
Lost some equipments.
Surveillance.
Can´t detect them as we should.
Few arts and crafts within.
DRCongo. Experimenting.
Bradley
Keynes
Feynman, R
Neil Kinnock

Feb 1st 08

The world all over again.

To make them quit.
Millions within or connected at a time.
19000 died. From wherever.
Sudden equipment failure.
A Tube disconnected.
Human mistake.
One of our great had to leave.
Mostly women. Of all times.
From the W-Hemisphere.
A Queen, princesses . . .
Venezuela.
Can throw their spiders twice as high.
The Avenger nonfunctional.
On a launching site.
Our vessels got cought in the web like flies.

Feb 2nd 08

Russia.
Kill us from within. Our Great. A Ghost.
Fierce response. Killed them from the middle ages
onwards. Tens of Thousands. Part of the Duma.
The European countries
closed their doors for good.
Kuala Lumpur. U.S. Base.
M. Proust
Balsac
A. Hitchcock
Australia wants out.
India too involved.
Japan. Arms from the mainland.
The better part of our pacific fleet.

The ship. Tried to repair it in haste.
Out.
Bush knew instantly. Some traitor within
from the beginning.
Islamabad.
Hisbollah
Egypt.
Karadzik getting stronger.
Jewish community in Russia.
The situation in Sudan - Chad.
Our fault. Lost the grip.
Kenya. Too obvious.
Japan. Buying up corporations.
The light is out.
(Means something).
Not a nation in the world that wouldn´t have tried.
To discover such technologies and watch the rivals
steal your own inventions to use them against you.
Our man Samoa islands
holds our Pacific population.
Coverdale, singer
R. Moffat. Missionary in Africa

Feb 3 08

Nairobi
Chad
The Russians started suddenly killing us from within.
One of their men blasted.
Within you: too young, too ignorant to have
power over life and death.
We killed young innocent in the wake,

those who had mere acquaintance to this
new thought, (arithm, calculations).
They still hold this deadly instrument.
Chad left bleeding, (for now).

Feb 4th 08

England. Pure thought done anew.
-(The decisions they make).
People is dying out of heart failure.
We crushed and crumbled your heart.
Run them through.
Some may have gotten away.

Feb 5th 08

Securing a strategic point we lost.
Few days ago.
SE Russia. Loss. Few hundred.
They have developed a most deadly weapon.
Elans brought to you.
The man responsible. Doctors. Scientists.
Our fleet. Fooled us over an area.
Activated suddenly.
Few thousand.
One of our men. Bradley.
Fierce response.
Innocent civilians.
Their hinges.
Their ceiling.
Wanted to create sth in here.
Takes more time. Have lost some strings.
(In hands, jaw,).

The pain inflicted upon one man.
One of our own. His thought somewhat erroneous.
Bradley. Gather his knowledge.
Elvis within. Leaked.
Strong in Ireland.
From Scotland.
Was head of our Russia army.
A division. Isolated.
Lost contact.
Wiped out.
We dropped it.
Few km. Radius.
They just perish.
Thousands of women lost individuals from within.
Anyone who is someone.

Feb 7th 08

Resilience.
Resources.
New device.
The search more sophisticated.
The aureole broke on the Greek.
Great many died.
USA on the doorstep.
Bush, Schwarzenegger
lose in spirit.
Kirgizstan
Ossetia
New elans.
Tolstoy
The magnetizer ruins their defence system

The simplest mechanism.
Any an Eifel would be proud.
Shields over strategic places
USA: their arsenal found.
Burma. Up in the mountains.
Russia. Their elite traced to Old Prussia, the Venetians.
Karadzik: out again
A platform.
Controlled from above.
Razor sharp.
The heart of the sea.
(Japanese invention)
Saline compounds
New possibilities.
The earth wrapped up again.
Net-points higher up than before.
(lost it at a point)
´A net´ can be thrown down from them.
Universal defense system.
No aggressors onto a neighbours independence.
An empire came tumbling.
Pernicious gift
Expatriate
Some too great to mention.
Beyond fear. If you´re guilty when caught.
The Gypsies. Have had to carry some load
for Old Europe. Weren't allowed to be with
at a point.
Fight and sustain.
Putin: Retaliate
Sarkozy: inclusive
Russia: has been fortifying herself.

Fed those within with fear.
Experimenting with your spleen,
The motoric system.
To make people lame.
No movement unless at will.
Give them the fright.
To persuade those in charge.
Not to wait for another to take over.
One of our great within L.M. Presley
thought to be a messenger.
Genealogy of the spirit

Feb 10th 08

Burma. Managed to shoot through
Our safety net. Their webs most deadly.
Armed men in the streets.
Combat.
Loss. Our: Few hundred.
New elans. Two of their men.
New material.
Light as if bounces back.
Ceramic can be used in combinations.
A probe on the launching site.
New detectors.
For outer space project.
Objects sliding from a field-point to a field-point.
Depriving other nations of the feeling of discovery.
Precious dangerous knowledge.
Heroin addicts used within.
Their sophisticated imagination,
their approach. Their color charge.

As if a force within
changes their appetite
,(on a helical level).
Just drain them out.
Cordova
Palermo
Lisbon
South-Central America links.
Immigrants from Africa
Problematic Dialectic
Transformation.
The product becomes the producer.
The simplest form of dialectics
Is in the form of induction-deduction.
Transformation
Rebirth
Becoming
System close / neither give nor take.
Dialectics in the synthetic form: repel - attract.

Feb 21st 08

The gravest cost
Yr head works as synchrotron.
Managed to increase the speed of particles.
Not a branch of science unaffected.
Changing eternity from a point in time to the present.
Argentina. Few hundred +
Belgium. The same.
Great explosion in Russia.
The Viscous incidence has turned out to be
a bad luck breakthrough.

Feb 13th 08

U.S. Pacific activity. Our vessels shot down
from above.
Devices hardly bigger than a basketball.
Spirals down too.
Controlled by a computer within.
The man known
The site known.
Response.
Lose in spirit.
Hours later they started to do the same.
So many of our great. One of our Ghosts
Within Al Pacino. Knew he was fated.
Took over the house.
Escape route.
Didn´t make it.
Few of their Ghosts lost branches.
One of the F.F: Jefferson.
Russia. Threw a bomb at us.
We dropped one as response.
Few hundred km south of Moscow.
Fellow countrymen.
Bowie. Beckham. Lost a lot.
Their spouses.
People in the spotlight.
Tourists, even.
Picked American entrepreneurs up
Wherever we found them.
All over Europe, Dubai, Tokyo, Shanghai.
U.S. Manila division. Out.
It wasn´t possible to kill within like that before.

The nations are entwined together
according to the taste of the times.
So many of our Great are/were within the States.
Most of our Ghosts too.
Take Hitchcock for example.
Half of his population is European.
The other half American.
Churchill. Partly inactive.
A.N.Whitehead. The same.
Los Angeles. Attack.
Dropped a few.
Surrender. Unilateral.
Prisoners released.
Weapons confiscated.
Securing Air-space.
New Elans.
Neil Kinnock +
(Within Al Pacino)
India Army.
Southernmost Russia.

Feb 14th 08

Get the job done

Feb 15th 08 0200 gmt

Day of Russia
They started killing us from within.
Didn´t know where.
First generation within.
17 thousand?

Among them our Ghosts, Churchill, A. Huxley.
Fright.
3-4th generation of their magnified, ghosts alike,
Lost much.
To cut their past from the present.
Until prisoners were released.
Surrender.
Weapons confiscated
Their equipment found.
Their communists.
Their aristocrats.
Magnified from Miloschevic.
The last three of their men. Made to enter.
Magnified.
Traitor within. Traced to Spain 17th c.
Few hundred ghosts dead.
Try to regroup them to gather their knowledge.
El Dorado
Venezuela.
Crazed tech. Throw up a web
which unfolds in the heights.
Scattering Archies. (air raid guns).
Our invincible fleet. U.S. post.
Many of our great.
Saved by seconds.
To make the shoe fit.
To come. Young rebels.
Ceramic equipment.
In the hands of the beholder.
Found. Confiscated.
Empty icons.
Pantocrati of the 19th and early 20th perished.

Took their men over.
(We can divide us more often).
So many of their children missing.
A group from us in each and every entelect
in Russia. We just run the show.
We hold their media,
Their army. Their ministries.
Institutions, corporations.
Intensive course. Russian.
Few conscience elans set free.
(From other countries than ours. Cuba fx).
Installing BTN. (British telecom network).
Each and every autonomous state.
Each and every individual within
Their men. Listed.
New world order.
Trans Ural.
Siberians?
Open Sesame.
Their underground arsenal; their pantry.
We are as alien beings. Strangers within.
Old and new tentacles.
Most of their projects cancelled.
Regroup their population.
by religion, race, demographics, geography.
Their directory.
Catalogue. Library. Database.
The Control room:
Foreign affairs.
Economic system. Banks.
Political structure.
Embassies.

Ships and submarines
Transport.
Airports.
Satellites.
Surveillance system.
Space programs.
Industries.
Their stock market.
Strange Dubai connection.
An Arabic front.
Their intelligences.
-(We hear, in some parts of the world
the least of regret. I.e. old curtain countries,
Natives trans-ural, in the extreme south).
Grouping some of their magnified together
To see who we can trust.
Tchetschnia. Who gave them arms?
China. Activity.
In the N-W.
Shanghai.
Venezuela-Colombia.
Cambodia. Rebel groups.
Russia. Mutiny.
Sabotage. Signalling.

Feb 18th 08

Thousands of those who were
Fighting and sustaining:
The commonwealth; parts of Europe.
Few thousand elans died from Koturbinski.
From Germany, France. House of Austria.

Russians. Magnified.
The doctor supposed to be responsible, executed.
Turned out to be tech failure.
Late 18th-19th c. He was to hold
The hierarchy for his nation.
Some of the autocratic states
are still with authority, dignity.
Sooth them by introducing our space project.
Found out how: a collective for (some of) your people.
Essence bound.
Now China.
Dismantled at a historic point.
Take over their men. Their new elans resp.
Regrouped with some of us within. (From a historic point).
Resistance. Those positively dangerous
Are sure to go. As will morally deranged.
Russia: oneness.
2 Cor 3:14
:The veil is done away
Apostles
Ministers
Ambassadors.
Confucian ritual
Summons sprits;
shamanism.
As orthodoxy: virtue,
order, harmony and stability.
Sentient: physical souls as opposed to
spiritual soul.
Analects
Regrouping ancient China. To change thought.
Hundreds of spirits died in the process.

Amongst them the founder of Dao.
We wanted to canonize from him.
The movement in your physical surroundings.
Seven story high.
So many intact at the same moment. (As pearls on a string).
Erroneous thought immediately afterwards.
A snowflake falling, a raindrop.
Everything loaded with meaning.
Those responsible executed.
(Two of your fellow countrymen).
Others died because of forgetfulness.
Since eve of spirits.
Forgot to measure their strength.
(Nano charge).
Came too strong within someone else.
Not enough space within, even.
Before. Greece.
A pioneer. Took part in creating the alphabet.
Others.
Emerald spectrum.
A magnetic plate, (a weapon), which folds out
From an instrument. Thrown through distances.
These writings as a referendum.
As if thought were spellbound.
Messages misinterpreted.
Tried to tell you about this religious sect
just minutes before.
So many incidents like that.
We cannot stop.
The old won´t influence us like that
unless at will.
To prevent the bad seed from falling.

The armed race took the flight
the day the world war II has ended.
Remnants of the Reich technology
Inspired conquering spirits.
Your family has suffered a lot.
Stabat mater.
On the soul.
Something only Bernard Higgins
knew how to do.
His way of thinking.
Those who constituted him cannot find it.
Things like that everywhere.
The war culminated in a fascism extreme.
Intellectuals, those who could think, executed.
The danger hour.
Shou Enlai. Magnified. The plane exploded on his way
to a meeting with Stalin. Perished.
Mao´s opponent. Possibly his successor.
Misinterpreted it some time ago.
Executed innocent people for it.
We have done them wrong.

Feb. 21st 08 Lunar Eclipse

Africa. Have been gathering their people
Throughout history. A homecoming.
The Arabic world, Old Egypt, the New World.
China. Age-old eternity.
Those we hantered with.
Their essence mixed with a
chemical agent from us.
Part of us moved within.

Russia. Rebellion.
Someone who was within
Stalin, Lenin, others.
Few hundred thousand attacked us
In their own houses.
The sea so silent.

Feb 23 08

Disarmament.
Weapons along the borders.
Chechnia. Türkmenistan.
Caucasus. Eastern territories.
Mansjuria.
Some loss.
Fire wall.
A mansion.
Inner Mongolia.
North of Gobi
Great Archies.
One of their men out.
Millions dead.
We lost something precious.
-(too bold an attempt).
P. Pot +
Action delete from within.
Ghosts, magnified, individuals.
Ghosts. Animosity of those who
constitute them.
No love. No feelings.
Pure thought. (Whatever that means).
Fearless.

Italy. France. Germany. Canada.
Need not only to sustain, but have actually
to fight with us.
Papacy. Spain. Portugal.
Their S-American interests.

Feb 25th 08

BBC on Russia: ´We are walking against thiefs´.
Henry VIII
Gone to the waters
H. Bequerel
Pompidou
J. Grimm

Feb 28th 08

A Coup.
New rulers
Imprisonment and execution
all over England.
Someone from Prince Charles.
16th Century.
´Knowledge driven Revolution’:
www.redirections.com
“The National institutes of health roadmap.
Space incubators,
Microbiome project, metagenomic approach.
Epigenetics: The study of changes in the
regulation of Gene activity.
The information operation (IO) roadmap:
information warfare.
Pentagon´s intention of gaining full spectrum dominance.

A key part in the military battle space.
Psychological manipulation.
Modify behaviour.
(From a 2003 pentagon document (not for release
to foreign nationals, including allies)."
Search. Feelings.
Hatred. Anger.
Anachrom recording
Anachrom display band
Argentina. One of their
Ghosts within Maradona +
Your Aura (as specs) inside a machine
Outside of. Sentients on specs.
The third ruling party.
These crucial years.
First someone in Adm. Nelson
Then someone in Ch. Dickens,
And now 16th c. Magnified.
Better Europe bonds, they say.
The economic man turn selfish.
The law of supply and demand, fx. means nothing.
More unemployment
Less GNP
More inflation
Companies turn their debts
into properties over night:
securities fraud.
Alexander Solzhenezin.
An Agent. Gone.
Fisher counterpart.
W. exile diploma. The lot.

March 1st 08

Girls of all times.
Edit. Longing.
Sexual activities.
As within older.
30 7

0

7 8
Nelson Mandela. In most of the houses.
Asia has to wait. Parts S-America.
Counter-coup.
(Conspiracy, they say now).
Their Churchill –line of thought-
Came through with Mandela.
The greatest fight for power.
As if two or three groups.
The older and the younger, iyw.
The Ceramic equipment.
The elder may have given Russia and America
The tech. To fight the conspirators.
The Crown. The world ruler.
Japan. New weaponry. Magnetic shield.
Their WWII man. 2nd generation.
Vessel down.
Revenge.
Counter-instrument at hand.
Chernobyl. A pernicious act?
The elder created sth. In Russia.
Fight from there.

Hid too well before.
Seized the opportunity.
Assassinations.
Lies. Deceit.
Treacherous people.
Around 1700c. England.
Political enemies, opposition.
Wiped out.
The same now.
The third party.
Andreas –line-, f.x.
The know-how doctor.
No obvious link to either party.
Too valuable a knowledge.
Try the neutral stand for your people.
A perfect world for executioners, warlords and spies.
Their week was enough.
We won the war.
Our sacrifices.
The invincible army.
Our houses: Beckham. J. Cleese. You.
Some were cheating colors.
One of their messengers.
Part Nietzsche.
The Rogues smoked out.
Scapegoats?
Trial.
War crimes.
Treason.
They shot at the same targets.
We had changed the coordination before.
Locked the keyboard.

Had them marked.
This Ghost-cabinet had to end.
They opposed the leaders.
A delirious attempt.
We gave them more and more power, and
as the tolls of the war increased:
they seized control.
They knotted young Icelanders
to themselves to see with them
And demand them.
China. Rebellion. Soldiers.

March 4th 08

Japan. Two new men. Space suit.
Cave experiment.
Crazed tech.
Traces to China.
Takeover. A generation lost.
Burma. Apparently stopped experimenting.
To love someone somewhere.
More at ease.
Have to unite a world.
Atmosphere leaking.
Each and every one of us.
We will lose the aura within the physical.
The soul IS the body.
China time.
Our people in India.
Indonesians.
House takeover.
Securing areas.

Eliminating opposition.
Our people from Hong Kong, Macao,
Shanghai,
Vietnam (French, American),
S-Korea (American).
Other.
Everything went hazelnut.
Japan. Tried to win over their houses.
Boobies. Will be annexed.
A. N. Whitehead +
A ghost from India +
Lebanon problems.
Can´t find our thoughts.
Empty spaceships.
Floating objects.
Scandinavians, Americans
given more responsibility in Russia.
We kill people on earth.
Gather ourselves in them.
The enemy does that too.
It blossom´d out of control.
First and second generation Russia.
9/10 male population in some regions
1/2 female, children.
First and second generation in China.
9/10 of the total population in some regions.
We still hold eternity.
Americans. Better lay low.
Kamikaze boobies.
Just waited for our house overtake
and blasted us.
Four of our Ghosts, magnified from wherever.

Who knows the coordinates?
Who knows about this country, this region?
Who can work on this machine?
Who knows how this device works?
How did we calculate?
Europe trying to break away.
Commonwealth countries.
India. Too involved, as of now.
Countries coming up
We didn´t have to worry about.
(Their thought relatively unharmed).
200 Icelanders +
People made to come.
Their thoughts right after.
Asking people to help.
Around Europe, America.
New people in charge.
(The old somewhat cut off).
Losing allies.
Canada. Holding back.
John Lennon +

March 7th 08

Japan.
Isotopes we hadn´t realized.
Fluid we hadn´t used in eyes.
Weapons, equipments confiscated.
Nations are a nought and void.
(The DSSC in English. Liturgical
Fragm. III Ch 9 p. 203. G. Vermes.)
India Muslim protector.

Indians as mercenaries in Dubai, Quatar.
Tech exp. Underground. Traces to Pakistan.
Your lines of movement
As part of the robotics.
Your organs, your bones, crushed and crumbled.
A tenth of 1st and 2nd generation in Russia gone.
A school of 1200 students.
50-60 within.
Things like that.
Prussia made to help.
Girls of all times.
Lose their sexual longings.
(Edit from within you).
We will just make them come, (love them),
at will.
Zimbabwe.
Their hallicunative.
Taste and smell. Addicts marked.
Any opposition fought with vehemence.
Social cleansing.
The Hybernians.
Those who tried the coup.
Trials. They locked some precious
information within the computer.
Never gave in.
´For the crows to plug´.

March 14th 08

A Barbpapa,
A jellyfish or
The gingerbread man

All in one.
Stomach. Outside of. Ulcer.
As light carrier. Vis: skin, fibres,
The spine. Bypass the regions.
Clamps.
Chemical agents.
The spine outside of.
Back injuries. For the lame.
Extracted as strings.
Internal tissues.
The stomach. Inside out.
As a bladder.
Through center of brain.
Asthma
Allergies
Individuals: ´I am the octopus.
When seven tentacles sleep,
One rests awake´.
Ceiling come tumbling.
Your nose. Cavity. Tissues.
Edit. Addicts. Won´t blow their noses so much.
Tibet.
Erroneous diplomacy
Central Asia.
Attack.
We lost few hundred.
Ahura Mazda had weapons hidden.
Bought from Pakistan?
Ghosts knotted together.
Einstein and S. Becket.
Both had lost individuals,
Magnified, Ghosts.

Stronger together.
Ghosts from Austria and Germany.
Hitler and ?
The same.
S-America.
Bógatá. Sites.
Critical points.
U.S. gangs. Crack.
Both Korea and Vietnam
Lost some Ghosts during the wars with the States.
The same during the world wars of course.
Your neighbours. Your family.
People you meet in the streets. Used.
Strings from you. Individuals attached.
Make them believe they´re somewhere else.
Experiments with the D/RNA. Protein base, gases.
You. Dressed in a film, through cuts, over the genital system,
Out back . . . Coated. More thorough search.
Individuals wound up on a ring. Millions in each.
Charge depletion. Response.
Few hundred thousand died in the process.
Dense light-bulbs thrown at you.
Individuals go through your system.
Yr. Aura. Scattered around you.
Individuals on each flake.

April 1st 08

Birthmarks.
Physical defects.
How we think/touch within during pregnancy.
Biomarkers.

Experiments within.
To please or displease.
To control in fact.
To feel loved.
Individuals bound together.
They never know who they feel for.
So many have lost their lovers.

April 5th 08

Plasticity.
Hard to detect.
Easy to hide.
Heat detectors.
Argentina. House takeover.
Sao Paolo.
Some districts.

April 9th 08

China. Tibet.
100 millions they say.
20-30000 children were over the blasts.
Some of them as old as history.
Between mountains high.
Few spirits.
Weapons confiscated.
Dharmsala. People from Dalai lama,
history and culture disappeared.
Warlords. Drugged.
Blind with hatred.
No imminent danger as of now.
Those responsible executed back in England.

As before. World throne.
Others had plans. Never helped.
Hid themselves, their relations, tried to eliminate
their enemies before.
16th century. Within Shakespeare.
Some Italians. Argentina governor made to go.
(One of theirs).
World trauma.
China. Abandoned building sites.
U.S. Virgin islands.
Spain. Cut off from S-Am.

April 14th 08

Türkmenistan.
The Kirghiz steppe.
A snake in the soup.
Arsenal broken into. A megastore.
Distributing arms.
Chemical factory. Bombed. Tremendous blast.
Sucking the atmosphere. Few km. Radius.
On the streets. Confrontation.
In search of their vaults.
Burma. Have evolved.
Weapon we hadn´t thought of.
Found the speed. The laws. Betrayal.
A security guard. Our loss: few thousand.
In the line of duty.
Children playing in the debris.
We are building a new highway over the pacific.
Someone from S-Arabia made both Bin Laden and Arafat.
The same Doctor made Castro, Che, Marley . . .

Harder to hide.
Russia. Too big a bite.
Messages misinterpreted.
Cocaine abusers.
Millions. Lose their ability to move.
Gerald Edelman (1924)
(See fx: world knowledge dialogue)
Neural Darwinism.
Brain based devices.
Conscience, he says, is a form of awareness.
It is a process. Continuous and changing
and has intentionality.
The brain as a Turing machine.
A refined dialectics.
The capable person: Ricoeur. Jean Pierre Changeux:
Workspace. Long-range horizontal interconnectivity.
Avoir, pouvoir, valoir.
Embedding. Manifold, graph, field
As an abandoned cocoon, nest.
Remoulding of man.

April 16th 08

Skin off. (Wrapped up).
Jaws. Horizontal.
Cut through center. Opened up.
For more quantities. The same with the ribs.
Those who bombed the chemical factory.
Found. Their own power brokers.
Strange accusations.
10 27
34

11 5
Rules, combinations, sequences.
There will always be an ´oppositionsgeist´.
Mozart without a Zauberflute.

April 22nd 08

Bolivia. Rebels. Addicts. Gangs.
Venezuela. Colombia.
Guatemala.
Honduras. Mel Zelaya. (President).
On a street level.
Tijuana. Ceylon.

April 24th 08

Kirgizstan. Karpata. Burma.
Russian and Japanese tech come together.
Those who came to power.
Their leader 19th c. Died within.
Pain inflicted upon. Their allies
in every cabinet. Their stronghold:
Churchill, Higgins. Someone within
Shakespeare, Darwin.
The great of the invincible army +
The Ghosts who were knotted together.
The one absorbed by the other.
Gases mixed with specs, spikes of light. Colliders.
Missile tech on the verge. U.S.A.
Distributing their loot from the megastore.
Know how we hid within their men.
Resistance in practically every country.
Their opposition party. Those who were working here

in the beginning. Some within Pavarotti. Verdi. Dali.
Some magnified in the States.
Three weeks ago. Drugs. Old Europe.
Raid. Ten Thousand.

April 26th 08

One of our great in S-America
got hit by one of our own.
Accident. Lost sight. Died.
Can´t trace his thoughts.
Lost inside knowledge.
Down Syndrome.
Eighth month of pregnancy.
Kerolen.
Liquid Energy
Sterilize from within.
Thousand at a time, or at will.

Mai 1st 08

S-W America.
Few hundred girls +
Were trying to double the quantity.
Russia. A thousand innocent killed.
Badly marked. Few fellow countrymen had to go.
Becket becoming strong.
Coup. Saw it coming.
Irish. (Part Sinn Fein was said).
Some in Crowley, Hitchcock.
The Irish in Australia, USA. Someone
high-up in England.
Had grouped possible opposition together.

One move. Hundreds. Or Thousands.
Parts of Becket.
How they have been treated throughout history
By their neighbours. To keep them under the throne.
Lost hundreds. One of our generals.
House in Russia. Practically empty.
Had given them an ultimatum earlier.
Double crossed them.

Mai 3 08

The East. Old Tartaria.
New weaponology.
A metaphor.
China. Took one house over.
Attack. Ghost India +
Two thousand.
Fierce response. A million?
Ghosts, magnified, children.
Lost some authority in China.
Their defence system has evolved.

Mai 5th 08

Venezuela. Occupied.
Man Colombia. Left wing core.
China. Preparing.
Rocket tech.
Implemented.
USA. Building arms. Someone who
worked here before. Response.
A. Schopenhauer.
Arbreu. (The author of El Sistema)

Those who have evolved
Decide the overall.
Syllables.
Consonant locks.
Intermediate reverberation.
As if the system refuses to iterate distinct sounds.
To change unwanted words, phrases.
To rephrase. To prevent or change thought.
A syntactic device.
Ghosts made anew from within. Individuals
´carried over´. Various reasons.
To be invisible within. As a security net.
Teeth. The densest light.
Vivify the roots.
So that we can sense our teeth again.
Some lost their lifelong companion.
A year ago.
Medical treatment for so many girls went wrong.
As if they dried up. Wrong chemical.
The spine. Outside of. For the lame.
Give back physical sense. Other reasons.
Days of awe. P. 142.
Compassion awakened on high.
For the remembrance of the early rising of Abraham.
(Ps. 69:2) The waters have come in even.
The thirteen qualities of compassion
The ten martyrs
The two covenants
Duties of the heart.
Maimonides
Th. Reid (1796+)
Holbach

Mai 11th 08

Burma. Disarmed.
Hardly a soul over areas, regions.
From a country to a country.
Hibs. Protectionists.
To prevent opposition of any kind,
any resistance, eliminate destructive forces within.
Means to an end. To secure control
To strengthen the family ties.
To control coldly. Completely.
To leave no traces
to prevent any assault
to break away enemy bonds.
Any unwanted thought.
To eliminate anyone who might incriminate.
Who might cause inconveniences.
Any demoralizer:
Social cleansing.
Inquisition
Prosecution
Shake´em down (and out).
The magnified and sanctified.
The great. Those who are capable of decision making.
Those who might run the show. Decimated.
Perishing by numbers.
Securing the position
strengthening the bonds.
Agony and pain.
Lifelong suffering.
Arbitrary aspect.
The rulers need no justification.

Show no weaknesses
no compromise.
We want to hear no complaint.
Feel no dislike
Hear of no contrary version.
Balance thought.
Things one does not think of.
Things that won´t cross your mind.
A third of the generation first
Has come to the waters.
69:2 Calvin
You´re under the lot.
The ultimate exile.
A reversed exodus.
Extreme reckoning.
Conscience elans from wherever.
All brought to one place.
To fight open or hidden resistance.
Your Sack. (Your lungs). Exploded.
Were used as a cavity. (Chemical agent).
Poisoning.
Many of those within or around died.
Late 16th c. Rule.
First of the Tudor kings executed.
Had a troubled past.
-(don´t work so fast).
-(and you didn´t trust me!).

Mai 12th 08

Those who came in their wake couldn´t keep it.
Had lost someone important the day before.

Didn´t have the army with them.
On we go in 1st person plural.
For whatever prize.
Call us protectionists for now.
Opening up some locks.
To have control over secret weapons.
Secure the mass.
Old Europe. Some of your elders in Denmark.
A group in Finland, Petersburg, Saragossa;
A group in England. An angler´s bait. We wanted the source.
Not so much effective polarity for now. In Europe at least.
The Swedish partook in this conspiracy too.
They betrayed them.
The Americans hantering with the eyes at home.. The helix.
Warriors. Die before giving in.

Mai 15th 08

Baffins Island.
Accident. Few hundred +
Up in flames..

Mai 16th 08

More invisibility
Government light.
69:2 Tchaicovsky.
Edgar A. Poe.
D.H. Lawrence
Rimbaud.
Jónas Hallgrímsson
R. Hesse
Göring

To the waters

Mai 19th 08

The States.
Took over their 50? Houses.
Rearranged.
The earth. A practical zoo.
A regular wasteland.
Consternation.
Accident within.
Fluid spelt on an instrument
You lost 15000 possibilities.
Mouth, toes, tissues, sack
To verify deeds. Sexual behaviour.
Oval fx. Strings. Workshop.
Military. (Used with the physical).
Mark distinct individuals.
Some great plan called off.
This instrument works with the sight.
One can ´call´ individuals by deeds.
The doctor who found out how to kill the brain
Completely. (practiced on a dog).
Within Andreas. Dead.
China. Boxers. At that time.
Found out how.
A woman entelect. In England.
Neutron bomb.
Radiant numbers.
Kerolen.
Actimide metal. 111
Isotrope+gas: fluid

Z bosom
The stock market. Capital piracy.
Suburbs. Veritable Zombies.
The polarity of power.
The House of York or the House of Lancaster
A Montague or a Capulet.
Whigs or Tories.
Eliminating opponents. Enemies. Traitors
Siding with the other.
Many of your countrymen who
worked within from the start have been called to the waters.
Bet on a wrong horse. Didn´t know who to trust.
Couldn´t come in straight.
Some tried to warn them.
They took the bait.
The fingers were done erroneously in the past.
To limit the individual according to deeds.
Some won´t point, others won´t marry . . .
Bonnygirls
Haemophilia.
Khrushchev
Brezhnev
Fed a group of individuals with distinct chemical drug
for medical purposes
few months ago.
Started dying suddenly.
Girls. Native Americans.
All too many died.
Estrogen substitute
(-gen base)
(-tein base)
Few story high.

All in one.
Every Ghost in one spirit.
A Disaster.
Wanted to try something new.
Millions dead.
Scorsese family wiped out.
Some from Monaco, Sicily. England.
Hallgerður Langbrók (Gerður)
Ketill Hængsson (Klængur)
Auður Djúpúðga
Gone to the waters.
As a punishment.
Nargos. Typhoon.
Didn´t prevent it.
Rather the opposite.
To cause an earthquake in China.
Create an eruption in Chile.
Plate base
Asthenosphere
Volcanic, seismic, orogenic.
Magma chamber,
Materials dissolve under pressure.
Lustre. (Reflected light from a mineral).
Metallic vitreous
Resinous silky
Pearly adamantine
Krishnamurti
Al Gore

Mai 26th 08

Since before the Norman conquest.

Scandinavia deep rooted.
13th c. Turks dumped hundreds of thousands
Of corpses into the black sea.
Early settlers in the Farao islands and Iceland.
Highly selective. Strong family relations.
Hid well throughout the ages.
Age-old ignorance. (To use a term from W. Blake).

June 2nd 08

Experiment within you with heat and cold in the wilderness.
1 as weaponry. For industrial purposes.
2 biology
3 medicine
Ganimedes
Neanderthal. Fake.
Composite.
One of the Dinosaurs. The same.
China.
Their oldest.
Sort of floating in space.
Their thoughts amount to nothing.
Their homecoming not so grandiose either.
Throughout the dark ages. As if they were fighting
an invisible army: the plagues.
Boulemi
The move about turned out to be catastrophic.

June 5th 2008

H. Purcell
Metternich
Human type form

0-field
Superspace
Those who were moved about died one by one
as did those they wanted to habit
They made some 600 thousand individuals on earth anew.
For greater quantities.
We don´t know how.
The soul turns radioactive when out of the box.
Attracts fundamentals from the atmosphere.
Sort of explodes.
Cannot reverse this process.
Various enzymes
Serotonin
Biomolecules
: a part of this mixture
Gene mutations?
Induces cell division?
Metabolism?
Breakdown?
Reverse pathway?
Causes chemical reaction.
Vital flow.
Experiments on prisoners,
An a living being. Try to target it.
Inhibitor?
Affects nucleic acids
Peptides, proteins
Transcription failure?
Translation failure?
A catalyst process when out of the box?
Lipids?
No energy storage?

Structure failure?
Signal failure?
A protein chain broken?
Antigen, (epitope)?
Antibody missing?
We, (of us within), attract when out of the box.
Have to regroup those within.
Cytosis?
Carriers (eg:cholesterol)?
Thirty possible reasons.
Essentials, chemicals, biomolecules.
Reactions . . .
Light.
Photosynthesis. Luminescence.
The set up of electrochemical charge
Energy boost?
Wavelength difference . . .
The nucleus explodes.
Nucleotides.
Irreversible
Tried to take out our specs.
Didn´t work
Bio half- time.
Quantizing some fundamentals.
A deliberate move.
People from the enemy line.
Families they didn´t like
People who made them (apparently) uncomfortable.
To eliminate thought from base-ground onwards.
Their immanent profit. Espionage
To be within their houses, magnified alike.
There were warning signs, as was said.

Wilh. Reich. Endowed with addicts.
Someone who was within Andreas Baader. Rote fraction.

Aug 20th 08

Madrid. Plane crash.
Sabotage
Within Saramanca.
Within Gaddafi
Within you.
Traitor. Supported Irish Catholics.
They´re sleeping on dynamite.
Eminem
Peres
Pompidou
Carlos. (The Jackal).
Didn´t bother to upheave him
Stephen Hendry (Billiard player)
Kostenista
Armenia.
Their hierarchy. Traced to Zarathustra.
J. Christ was within.
Ancient Hindu
Part Illiria, Greece.
Türkmenistan. The plane crash. Probable sabotage.
Khmer Rouge link by the time of Balukistan.
(Bangladesh and Pakistan resp)
Old Armenians within. Link we hadn´t found.
The Russian Resistance. Torpedo. Telecom system, drawings.
Tried a coup two? Weeks ago in one of the men. We lost a few.
Mostly from India.
Someone within notified them.

Turned ugly.
Mass hypnosis
Matchmaker
Toxic waste over waste areas
Eats the atmosphere up
No force field works on that.
Worse than ever.
By mistake: experimenting with so dangerous chemicals.
Deliberately, or so to speak: the war.
Germs.
Polluting reserves, areas.
Spreading fast because of man
The Rosicrux crusaders slaughtered the Armenians at a point.
France. The Vallons in control now

Aug 26th 08

Japan. Occupied. 40? Houses. (Entelects).
Some of them look strangely indifferent.
Others strangely ignorant.
As if some of them don´t care for their lives.
Nothing to lose any more.
Their code of honor.
No apparent weaponry.
Surely hidden somewhere.
BTS inserted
5th generation since this new invention.
New hardware.
New Ghost. From England.
Helped him over it.
Twice the woe.
Securing air-space.

Detectors, sight, from the heights), adjusted.
War crimes. Addicts. The lot.
A million addicts marked in just few hours.
Coke mainly via Peru from Colombia
W. Disney was within in Japan. Others.
Light shed. Last of the centerfolds.
Burma. Building up an army.
Nebraska. A site since the days.
The Spanish speaking in the States.
Arabs/Muslims in the States.
So few within. Killed after the TT attack.
NASA. One of the first trips to the moon was staged.
Warsaw. U.S. link. Old Giant letterers in the new world.
Bad seed infects thought.
Endangers others.
Beat the form of the tree
Erik of Pommern.
Erdogan

Aug 29th 08

N-Korea. Attack.
Quite a few Russians within. Mongolia link.
Securing houses.
N-E China blasted at the same time.
Japan. 19th c. connection Germany.
Goethe had been within.
The instrument moved over Afghanistan
Addicts. Rulers imprisoned.
If the magnified individual uses. Dismantled.
They shot last of their mountain lions these last days.
Extinct in that part of the world.

Sept 7th 08

Off the coast of England
A Catastrophic accident.
Coup. Here.
Resistance world over.
Thousands of young addicts. Shameless and angry.
This form of resistance. Two incidences these last months.
The Factory before. Sabotage.
They suddenly came to be in so many places:
think they´re on the top of it.
Saw the whole thing coming.
A dint of lies, deceit.
Warned them, even.
They´re gonna have to think twice.

Sept 14th 08

Voice assimilator
Intercooler
New detectors
Use UV light.
Fire wall. Shields.
Indonesia.
Portable spiders, powerful lasers.
Fear of Nuclear.
The cities began whispering.
8 millions marked. Sulawi, Java, Sumatra.
Our loss. Many from Australia.
Site found.
Civilians, age-old Arabic, Indian.
In fact all of the Muslim world.

Traitor within Negroponte.
The Aeroflot plane in Parma.
Explosives.
May have prevented similar incidence recently.
Resistance in all the corners of the world
Negroponte

Sept 16th 08

China.
Heavy combat in the north.
Took over few houses.
Lost few hundred. Tactical mistake.
Rather die in combat than fade away in prison.

Sept 18th 08

Akasia
Parts of the Red army within.
Armenia, Byzans
50 old spirits. Few of them listed before.
Some had lost a branch before.
Zoroastrians,
Old Greece. Some only existent there.
Indian. From eve of eternity. Old Hindu gods.
Their influence tremendous.
Some thought to be gone. Had branches in houses
we couldn´t touch.
Fought the war. Throughout Evrasia, The Peninsula, Indonesia.
Someone who has always seen within.
The Center point of Everasia.
Few branches from old Russia.
Kirgizstan. Xinjang.

Extreme S-W China
76°E41°N
150 km NE of Pik Revolucii (6974)
Pik Kommunizma (7495)
On the Scandinavian Revolt (April?)
One of our four rivers gone. Norway fons.
A golden bough myth.

Sept 20th 08

Machine failure
Over Arabia.
Short Circuit. A device erroneously done.
Thousands.
Later. Here. Were running people through.
A mistake. Those who were working within had to go.
A woman held responsible. The same.
Hatred towards the rulers. Not tolerated. Perish.
30-40 of your fellow countrymen.
Most of them from an old family. Had to go.

Sept 25th 08

A Device to detect nitrate bombs.
Portable.
Activated by sensors.
China
Took over another house.
Central mastermind.
Their ceiling tumbled last spring.
Hardly a history.
Our Rule.
The Major. His rigour wasn´t enough.

Another. Lasted a day.
Someone who was within Shakespeare seized control.

Sept 27th 08

Over the arctic
A blast.
Almost simultaneously one of our ships exploded.
A Cruiser. 20-30 within.
A surveillance system on the Cape of Good Hope.
New Crew. Some of us were here during the invincibles.
Couldn´t find us.
New people in every port.
We had to cause chaos.
The Arctic turbulence.
The Ship.
The Good Hope base.
A part of the scheme.
Of those infected. Their DNA doesn´t bind.
The C-receptor.
A peptide.
Essence of the very soul dead.; still born.
The ´youth´ counterfeit. Part of it from us.
Fourteenth century. Blazed through the north coast
of France, Belgium.
Could hide within after war compensation.
Few of your countrymen had to go.
Sold us out. Were working for the wrong people.

Sept 28th 08

The Cabinet of yesterday quite simply gone.
Friday night. A tornado of youth from our own backyard.

200 from your own country. Calculate: times population.
Pearly drug. Labo drug. On a gas cloud.
On earth. Pills mixed w. heroin. Fearless. Crazed.
Killed those they didn´t like. Most of the board.
Few old with them.
Know all the tricks of the trade.
Find their likings. Recruit them.
Fooled security.
In disguise.
Had been studying us all along.
Where we were at, who´s within . . .
Someone within the ship planted it.
No bigger than a box of matches.
´Sacrificed´ himself , as they themselves said, to
distract us, to gain time . . .
Somehow truth lost its meaning in their minds.
They have seen it all, been to all corners of the world.
They figured out how to hide their addiction,
Shielding each other from within.
They killed tens of thousands before we could stop them.
Betrayed our trust.
Most of them were within ghosts before hell broke loose.
No witnesses any more.
With a wit and some intention
Anyone can fool cameras and surveillance system.
Gone the great, the lonely, the meek
Gone the heart of the planet.

Sept 30th 08

Crystal
Zeolyte Structures

X

The peninsula, Himalayas.
Lost some regions over there these last weeks.
Burma. Gathered their strength. New weaponry.
On the defensive. Two houses found.
Muslims admiring the tech.
China. Before. Had developed giant magnets.
Age-old Indian princess.
Homosexuality throughout the generations.
Gopi girls of all times.
Lang Lang (Chinese pianist)

Oct. 1st 08

Topological density
Electrostatic surface
Target temperature
Zen.
Bali
Up in flames
Archies down. Labs found. Street combat.
Lights went out throughout the islands.
People had gathered in barracks.
Didn´t hold much. Burning.
The red cross on the schene.
Their war cabinet found
Warehouses. (Arsenals).
Traced to the ancients. Hindu.
Bad company list
Two distinct companies.
Their roots and branches so entwined.
For greater quantities within.

Brought to you or some other house.
To verify deeds. Espionage.
We are 100x smaller within.
Most of their, (our, iyw), men made before June 2008
might perish as well. Thousands of families.
Politicians, scientists, media people, actors, artists,
People of social importance, people in the spot light,
Royal houses, great families.

Oct 3 08

The chair too hot to handle
Cleanse ourselves from adversaries, rebels
Opposition, sinners alike.
Few hundred of our entelects on earth.
Alive but not living.
630000 of our magnified
Not even embraced by the atmosphere.
Receptors. No self, really.
New Ghost. English-
Australian a month ago.
Two-three months and their ´first´ is
already also third gen. within.
Every knowledge on earth gathered in one place
(to try to find the cure)
We have to get former size within during upheaval.
Tried to feed, (prisoners in fact), with peptides.
Couldn´t communicate.
Different wavelengths
Catalysts
The right peptide at the right time.
The right mixture.

Timing.
Was alive for some seconds.
17 millions connected to you at the same time.
Winston Churchill worked in the spirit of sacrifice.
Some within him decided on this master plan.
They thought the Swedes how to work with
great quantities within.
As most of their allies. Others just by command.
So many things they tried that didn´t work.
So many victims.
So many handicapped.
Some acted in good faith.
Didn´t cross their mind that none of them
would not be upheaved.
Last of the battles in China.
Sites. Laboratories.
Want their megastore.
Can´t have them distributing
Arms to rebel groups.
Move prisoners about.
Up the hierarchy into adversaries.
Tried new dietary.
Peptides, serotonin,
Learn their signals intelligence from up above

Oct 10th 08

Our ship down. Asia.
They hold a nitrogen-device.
It sort of jumps over and burns
In a distance.
New people in charge.

Seized the opportunity.
5000 of your Icelanders participated.
Took over two houses.
Two of the lesser giants iyw.
To upheave in haste if needed.
Scandinavians, Canadians, Americans. Recruited.
Han country relatively safe.
Mansjuria relatively safe.
Some magnified hiding.
Perpetual danger.
Petrified.
Their first born (India).
Upheaved one of our lesser.
Lost one wing immediately.
Intensive.

Cleanse.Oct 11th 08

In fact everyone who could lend a hand.
To get away with the English.
To free the world.
Connection to the Scandinavians
Strength in N-America.
Their inborn will to power.
We suddenly had their own.
Took their own houses over
Prisons opened for conscience prisoners
The world dominion of the British empire is over.
London stock market.
They were openly looting other nations' riches.
Their ′transparency′.
We had become an integral part of their war-machine

As 2nd or 3d generation within their collective.
A vast minority has oppressed the nations.
They just lost their -elegance-
sometime in the past.
These lofty promises.
We isolated their great,
Synchronized our movements
Some of their rulers were victims of circumstances.
To secure peace
These last days. Hundreds of thousands from England.
From their great onwards.
To silence them once and for all.
They just lost their grip on you.
Regroup ourselves
in each of the Scandic countries
in every continent.
Help them over it.
These constitute, as 2nd generation within,
one of your fellow countrymen.
Helped him over it,
He in turn moved around the globe
To be within their entelects, magnified.
England subjected.
Their pilots given leave.
Old Europe retrieved.
Colonies of old in various places.
Rearranging world order.
So many of us took refuge
When the storm broke loose.
Leave all quarrel behind.
Approach the problem differently.
Over Saxon hills

Over home.
New relations
New bonds and old ones retrieved.
Bad company list
Esso and BP
Visa
Nikon
Siemens
Sony
Homemade nuclear bombs.
The simplest technique.
Explosion. On innocent.
Traced to Bagdad.
Rebels. Addicts. Few thousand.
Von Humbolt
Noreddin
Absalon (1201+)
Melanchton
The entelects all in one
India governs the better part of Asia major
We merely supervise
Charlemagne and his three swords
Curt- V-Francs
Almachia E-francs
Dyrumdali Lothringen
Secret ladders up in the hierarchy (and down).
Christology:
Indivisible unity
Divine nature
Hypostasis
Resurrection.
Remote Russia

Attack.
Simplest air-raid.
House out.
Few thousand.
One of our greatest
Had few hundred thousand marked.
All of them in one go.
Sjamanism
Krsna, Buddha
Some old sects
Poisoning our thoughts
Feel it coming from above.
Distract us with the attack
On Christian minorities, Iraq fx
No boiling pot as of jet.
Not long range.
No giant magnets.
Some want to take the ceiling off. Period.
Compensation.
Hundreds of thousands English gone these last days.
Their army from the top to the bottom.
We were partly punishing them for what they had done to the enemy.
Another house in great danger . . .
The shields took the blow
Few hundred Icelanders gone.
They let us migrate over there.
They will try to let us hate each other.
Bring back the polarity of power,
As they did the English.
The elder influence so greatly.
Lost some precious knowledge monetaire.

The commonwealth had a stronghold
In the India subcontinent.
Some of ′our′ eldest (through Byzans) closed their windows.
They have been there for thousands of years.
Know every little stone in the landscape.

Oct 21st 08

A kind of a league of Nations established.
To minister the planet.
Australia subjected.
Cryptic information.
Their observatory,
their turbines, factories.
Few of their abandoned programs,
labs, Kerolen refinery.
They killed each other until we eyed a chance.
With their elegance gone,
Their greed and arrogance shone through.
Break every tie they had with other nations.
Wherever their ships have sailed.
New Ghost, Russia
Armenia. They found their Greeks again.
Odessa, Tyrol
Millions from India within.
They have to trust us.
Just one move if they try something.
The last of the English still hiding
In Kenya, Ghana, Hong Kong, Canada, India.
A list.
The northernmost coast of France,
The Netherlands. Gaelic link

The ministry of Defence. Our responsibility.
20000 Icelanders come home from Europe
From Europa, from Norwege, Hull, Grimsby, Le Havre
Refreshen our bonds.
Profit from their technology.
Tiamin
Tuberculosis

October 24th 08

We still have a stronghold.
Hold specks from every human being
On this planet.
A kind of a stalemate.
Continue with our daily work.
To secure peace.
Cleanse the world from sinners, criminals alike.
They announced it on some pirate station.
If they would find out how to take down the paralyser.
(Paralyzes people over vast areas. To pick them up).
They need someone for a suicidal mission to get it down.
The eastern parts of the Himalayas.
Cannot find their war cabinet.
This equipment too important.
Imbalances world order in their favor.
They are sure to have what the English had.
Nuclear, fx
Our sensors, detectors, our surveillance system
works on ground. Not a missile moved unless
we know of.
They have a great army. We, reversely, are
Inexperienced in combat.

They have evolved their telecom sys.
Some sites the English abandoned.
Might be working all throughout Asia in fact.
Beyond diplomacy.
Close call. They tried to take over the army.
Their fleet changed base at our own accord.
Weapons hidden underground.
Exposed. Had a secret program.
Their generals.
Military officials.
Managed to persuade them.
They will go by numbers.
Europe. Have to make us new ones.
For India division. Germany. More involved.
We were evacuating areas. Too close a call.
The Maya culture.
Confrontation from India.
Few old from China, who lost their place w. the Inks
During the colonisation.

Oct 26th 08

Our allies. Vernacular.
Codes.
Few thousand out.
Tantalizing effort.
Links. Since the formation of the Commonwealth.
Netherlands. A few who were there in the beginning.
Some from Humbholt, another Ghost we decided to trust.
Gave them to be within. Asked them for responsibility in fact.
Resistance never sleeps. Organized few could recruit
a million in a day.

Be it an image of a person, calculations, drawings,
or secretive documents, a chemical structure or any a deed, we hurt suspects by impressing images upon them. (by millions). If they recognize an image they let out charge.
The syntactive device works similarly.

Oct 27th 08

New ghost, China.
Eliminate polarity. Communist crimes.
Give back to the people. Secure stability.
To simplify the situation in China:
seven dynasties. To bring them all in one.
Couldn´t bring them all together.
New rebel parties. Fractions. Their isolationalism.
Mongols.
Mediaeval confrontation w. the Germans.
Some of their NS only existent there. (Cut off
From Europe).
Mujahaddins. Parts in Turkey.
The young. Our children. Were within the elder.
Out of necessity. Won´t so much anymore.
Addicts used them. Their responsibility was overwhelming.
To act as an old individual, listen to decisions making,
influence and be influenced upon. Used as nannies for the mothers, as messengers, misled by stronger individuals, be inflicted by
The deeds of the elder.
Baudelaire +
Verlaine +
Laxdælu-Bolli +
A remote control

Now the scelette,
Now the organs,
Now the genitals.
Quite simple in fact

Oct 28th 08

Insurgents on the borders of Afghanistan
and Pakistan.
A spirit in India. Dismantled.
Russia (Ch.H. Pearson, first ed. 1859, Frank
Cass & company, 1970)
Written during the grand reform
The epigram within.
P. 11 The old character of Germano-Chinese civilisation.
On Caucasus, p. 137: the mountains a resting place of the desperate.
139 the dread of divine jealousy.
144 To see the proof of divine right in success.
151 the grand Hebrew commonwealth which Calvin tried to realize.
155 Silently and almost unconsciously the people, youngest born of time, has become the Empire.
Peter the Great
Bixdehude (8th c. ghost. Some from King Albrecht within.)

Oct 31st 08

A powerhouse exploded.
Sabotage
Spanish, English.
Georgia. Rebels.
Lost a few.

Erroneous diplomacy.
Our allies. Europe.
Not with us as we expected to.
Synchronizing our move. Entelects.
Most of them with a remote. A part of the cortex,
The hands etc.

Nov 2nd 08

Few Norsemen. Didn´t help.
Rather the opposite.
Swedes. The same.
The Australians in Japan.
Planning. (impregnated by the English within).
Drive them out of Japan. They won´t train their
Rebels from there.
New equipment.
At any point-space. Aim from within.
Works with our lense system.
South of Goby. Rebels. Someone high up.
Tens of thousands.
New Ghost. Germany.
Homosexuals.
The elder within persuade the younger to participate
in their doings.
Quite a few molesting their genitals in some pervasive act.

Nov 2nd 08

One of the hinges fell down. (A platform).
Ahasverus.
Followers of Lao Tse.
New rule

Under his rule; millions in one month.
Shouldn´t take long to run you through.
One of the great died. Was working in
your reciprocal. Boobied you.
Recuperating.
New recruits.
The overall is more or less
the same now.
They came with the German Ghost.
Blind to dangers.
Inserted them manually.
Few Ghosts made to go.
Humbholt +
The Hugenottes helped them.
The French. We had decided to trust them.
Traced to the Medici.
What did they think?
That we didn´t have them marked?
That we couldn´t edit them unless from you?
Some Icelanders helped them.
An old spirit from India +
Was fighting us on all frontiers.
A micro softense
Inserted in the plasma.
The hierarchy becomes transparent.
When a spirit dies; his children left unattended.
Chose a new leader. A woman. Partly in Canada,
Norway, Iceland
On India.
Has evolved
A direct threat.
Persuasion.

Nov 8th 08

Addicted youth
giving countries to friends.
Slave trade
The worst of vices.
To bring end of history.
Someone wanted a lion,
Another wanted to eat a penguin.
Their harem.
A Couple. The woman in all four of his wives.
No boundaries.
You´re the core of it.
A dreaded machine.
They can command as long
as there is someone to obey.
The young. They killed thousands from England.
3000 from Russia
7-8000 from India
200 from U.S.A.
So easy to kill people from within.
Hartford +
The cure found for these 630000, (200 of them Ghosts to be).
Dehydration, they labelled it.
Helped one Entelect over it. Needed another for Hartford.
Dropped a small nuclear on an airbase. W-Coast, U.S.A.
Ramaputra
Their weapons confiscated
New equipment. It can lift itself in the altitudes and shoot an any direction
They were preparing for the final battle.
Beat them by the hour.

Nov 11th 08

The technology we received from the Indians.
Force carriers
Tinier movement
In the brain.
More efficiency.
Wrapped the world up in higher altitudes.
Secured our houses
Some came from Norway with a secure pass.
Took out one of our leaders
Icelandic traitors. Someone deep-rooted in Norway.
The hinges are more fragile
The instruments attached have to be lighter.
Coordinated effort. Started in India. Few dead.
Blew up a house in Venezuela.
Drug of fearless.
Give it to them.
Don´t ask.
The seed of unamnity.
Some deadly weapons
Circulating underground
Loot from Megastores.
Some the enemy at that time
managed to hide. Smart bombs,
Hand grenades, Guns, etc.
Someone leaked info
Surveillance system failed.
New magnetic force.
Restrain the fundamentals with.
a secondary force
World veto

Managed to integrate with the magnified.
Meeting at the world council.
  High commissioner
  Committee
Departments
Distributing
Shared responsibility
Future prospects
Isopropanol
Similar to Camphor
Makes you careless.
A sort of witchcraft
It sort of puts you off guard.
Lose judgement.
Tried to put the machine at fault.
Idleness of the spirit
Loses part interest.
Attila the Hun
Diderot
Esterhase
Moliere

Nov 14th 08

Diamond sharp (hot)
Chrystal sharp (cold)
To cut nanoparticles fx.
Space incubator
2000 km/pr. Sec
Nemehia, Noah
Cytoblasts
Imps

Manifold affinity
Tolstoy
Lost most parts during the war.
The Chinese lost a thousand years of history.
Try to retrieve it from their neighbours.
Found Milton in hiding.
Shakespeare, King Arthur, Vulfila within
To name a few. Lost his ground during the war.
In India, In Russia.
Their Pirate-Ghost within. Someone else named Hennessey.
One of their generals. Attacked L.A. last winter.
Killed his Russia population.
When the Vikings sailed up the Thames,
The English had attacked Denmark.
Their historicism one sided.
Nero
Constantine
Cedrik
Ming Ivan Gavrilov
Diego Rivera
General Custer
Lightfoot. Sioux.
Within a Spaniard.
Mythology attached.
Within Jefferson too.

Nov 20th 08

Europeans
Attacked one house.
Lost one of our great.
The papacy.

Dominicans
San Francisco.
Tens of thousands
homosexuals
prostitutes
addicts.
Old opium dens.
The ´coolies´.
Remnants from Chinese slavery.
Poland. Group of Swedes within.
Vasa Dynasty
Old Hindu God.
Fought the English.
Organized himself from Burma
Catherine the Great. + (during the war)
Marie Therese +
69 :2 Joan of Arc

Nov 22nd 08

Tried to overthrow the rule.
An old war cabinet.
Few old families in Europe.
We lost some precious.
Riot in one of the houses.
Had marked us beforehand.
Rangoon.
Explosion.
Plastic explosive.
Pali Empire.
Fight in Dagestan.
Few hundred thousand +

Old Chinese and Hindus within.
Bagdad. Bus. Explosion.
13 women on their way to work.
Killed a million in the wake.
Young boys organizing.
Morally strong women have to take lead.
Mumbai attack.
Tens of thousands.
The English were favoured by the Greeks.

Nov 30th 08

New colour.
Ecre blue. Aids.
Deep inside the rectum.
A fluid.
Our underground system.
No maps whatsoever in some parts of the world.
Tunnels no one knew about.
A haunted Church.
In Russia. Under an orthodox church.
Sometime in the past the tunnels exploded.
A University Campus in the States.
Underground world
The cabin in the forest.
They came up from underground.
Boobies.
Tatchekistan. Te nuclear bombs they stole
lies hidden somewhere underground.
If well placed they might kill hundreds of thousands.
Meeting places for addicts rebels alike. (Gas compounds.
Labo drugs).

The greatest caves have formed due to our use of
minerals and oil.
Hard material has to be ′sterilized′. Impossible to tunnel the sea.
It cannot be sterilized. So alive. Constantly moving.
Turbulence, resistance, pressure. Gravitational harmony. Force field. Field equilibrium.
Meteors. ′Dead′ material. Sterilized.
The surveillance system. A tedious job to monitor the screen.
Constantly trying to find strings from age-old history in the countries who got worst hit. China, Burma, Russia, Venezuela, U.S.A., the Arabic world.
They all lost a wheel of nature.
Most of our old Icelandic families run through Scandinavia.
Try to establish contact with those who want better
for the world.
A Viking colony in the east.
A Viking colony on the west coast.
As part of the Holy Roman empire.
As part of the Byzans empire.
A Suisse. The oldest families in Europe within.
Pervasive sexists. An activity which has been labeled
Sadomasochistic relationship. The younger within,
loyal to the older, after years of subversive training,
infuse or accept pain, while the older actually commit them.
Lesser evil shelters worse evil.
Another example. An individual considered to be insane.
The very soul within him uses him to distract her own doings.
The same with addiction, drunkenness, tobacco smoking even.
Intoxication in one place causes a movement in the other.
Carelessness. By product malignant thoughts.
W. Scott

Des 1st 08

Conspiracy.
One of our Ghosts killed.
Some 20 houses unsecured.
U.S.A. England. 200000?
Lisbon, Milano, Plymmoth
In India. Experimenting behind our back.

Des 8th 08

Micro-threads to bring back the Chinese ceiling.
(Sort of floating in space, as was said).
So few strings to earth. No communication.
Seems to have worked.
A Ghost we didn´t know about.
The Ghost Shangai was within Mao, fx.
If the English had found him. Mafia.
Tens of thousands of Europeans within.
Addicts.
Disappeared in the midst of the war.
Was anchored somewhere in hiding.
Finding strings of families. Hope to minimize
their loss. Patience and time. Where they
are anchored on ground. Find ties.
In India, Mongolia, Japan, Korea, the peninsula,
The U.S. even.
Critical eyes.
Puerto Rico link
Last move in the political landscape.
19th c. English tax collector. (Opium).
Tunnels from the Vietnam war.

Des 9th 08

9 months since the orient lost most of their ghosts.
Practically without a minister of the flaming fire.
Up until now, Humbholt didn´t cooperate as was expected.
The Australian died on the first day because of dehydration.
To bind those trustworthy.
Find missing loved ones.
From the root to the periphery
Aerial weapons throughout history.
Spears, swords, lashes. Noradrenalin

Des 10th 08

Zacharias, others.
Tried to cause unamnity amongst the English.
They thought they were too impertinent.
Made too many mistakes.
The battle for Stalingrad during WWII.
The Judes and the Pure turned against
The Germans.
A tiny instrument distributed a year ago.
(Few micros). Called a splinter. The youth
made to give it back. Some had adjusted it.
Practiced roll play within.
The acupoints. Some figured out how
to use them for themselves.
A labo-drug they distributed for personal use.
A year´s supply hither and thither.

Des 12th 08

Finnmark.

An old Lapp. (5000+) Couldn´t upheave him.
Was supposed to unify the northernmost parts of the world.
It started with the killing of an American soldier.
Our Arabic governor resigned.
Salman Rushdie saw it coming
France. Building site. Persuasion.
Germany. The same.
Static energy.
Aerodynamics.
Sound. Smaller. Only 16 nanos plus a few to fasten the thread.
Menasse
Noreddin
Gutenberg
Grundtvig
Tycho Brahe
Niels Bohr
Esterhasy

Jan 4th 09

46000 Jews in prison or dead.
Young and old.
Big ones.
They won´t transgress us like that.

January 10th 09

Karadzic. Strong Muslim population.
The mines of Salomon. Location known.
Loot in fact.
The great plague. The Jews told their own people what do to and what not to do.

Jan 18th 09

Kasaars.
Seljuq Empire.
Bagdad. 32000+ mostly civilians.
Underground. Tunnels fell apart.
Some who took part in the Mumbai incident.
Pirates from Somalia, rebels from Iran.
The explosion was too strong.
We are partly responsible for the Israeli invasion
into Gaza. Could have forced them to stop at any time.
We´re old in Innsbruch, Tier, Inverness,
west of the Rhine. The Rhinelands.
Fontainbleu
M. Planck
Ataturk
Glanular product extracted differently. Baser.
More efficient.
Artificial lungs. (A bag).
Burgunds. (Sigismund family, Faefnir killers)
The last to abandon NS machine.
The 7th c. king didn´t work.
Zeno chosen to supervise. Was in a mediaeval sultan.
Cassiopeia. Her murderer found.
The lost church. Not Jesus. No cross. Europe.
A new Ghost.
-(Strange to be old in time and first person again.
As if the
ceiling comes over you).
New friendship
New world.
Abraham. So strong within both the Hebrews and in Arabia.

Feb 8th 08

Eiríkur rauði. disappeared into the sea.
Egill Skallagrímsson. Shot.
Bríet Bjarnhéðinsdóttir. Killed by the English.
Jónas Hallgrímsson. Boulemi.
Guðmundur Kamban. (The Danes only just uplifted him for a trial).
Þorgeir Ljósvetningagoði. To the sea.
Brynhildur frá Snæfellsnesi. Killed during attack.
Jón Sveinsson. Deep rooted in Germany
Gunnar Gunnarsson. "
Jón Leifs. "
Mikines. "
Knud Hamsun. "
Frau Hamsun. "
St. Patrek To the sea.
Ólafur Helgi. To the sea.
2, 3 million of Jews who died in concentration camps.
Those who perished by the nuclear bombings.
Hallgerður
Auður Djúpauðga
Ketill Hængsson
Imra Gruse
Eva Braun
Queen Mary Stuart
Sir Lancelot
Those came back to name a few.

Feb 10th 09

Troy. The siege. Massacre.
The library in Alexandria.

Incendie by the Romans.
New life-form.
We created a planctum.
The Giant carcinogen (cancer) extinct.
A Ghost. Saxons, Judes and Romans within.
Some 20.000 within. Big at that time.
Brought back by you. Early fourth century.
Made for Christianity. Killed in England.
Rommel. A Ghost. Brought back.
Some need a world of their own. The eldest.
Their sight ruined. Gazing at the constellations
For thousands of years. Some don´t have sense in their fingers.
Most feel less.
Entelects:
William Tyndale. They burned him for witchcraft.
Didn´t upheave him.
Calvin: Died in a war
Isherman: Killed by the NS
Thracians. Phoenicians. Spartans. Our forefathers.
The philosopher Pythagoras brought back.
We now can upheave an individual with a thousand units.
Up ten generations in just one night.

Feb 11th 09

Terms of trade
Unfavourable import pattern
Target price. Provide intervention.
Tied price.
Regulations
  Directives
    Decisions

Recommendations and opinions.
Public companies.
Powers and obligations
Monetary standard
World resources
National resources
Indices
An undergrounded economy.
(see fx. The fontana economic history, ed C.M. Cipolla):
Idyllic stability
Managed money
:increased state monarchy
Gosplan
Functional nature
Short term plan
Long term plan
Gestation period
Control figures.
Market equilibrium
Strumilin: Steady decrease in market relations
>system of priorities> economise
Initial targets. Requirements:
Norms of input use per unit of output.
Plan
Implementation
control

Feb 12th 09

Kidneys. Left to right.
The heart. Recreated.
Fingers, the tongue, finer tissues.

Wiesbaden
The steppes east of Volga.
Been there since before our era.
Relatively peaceful for a thousand years.
Actually in millions on earth right now.
So many addicts, the English had thrown out
Came back down. Our rulers, Ómar, Vignir,
Vilmundur, (Argentina),
Gunnar Gunnarsson, (China). 20.000 of them
from Iceland alone.
Luckily for us others, we were upheaved with them
High up in the hierarchy. (Family relations).
Facto grege
(the herd has been assembled).

Feb 15th 09

ArXiv: 0903.1228
Various
An extended network
Gel-forming mixture
Bond loops
Aggregation
ArXiv: 0903.1199
Th. Alarcon, H.J. Jensen
Spreading a complex network
Percolation
Degree distributions
Exponential, scale free.
ArXiv: 0903.0740
K. Ziegler
Mono and bilayer grephene.

Fluctuating gap>random staggered potential
Continuous symmetry
Two-particle Green´s function>mass less fermion mode
Dominated within a loop expansion
By small loops>diffusion
Neutrality points
Density of states
Insulating behaviour.
ArXiv: 0903.0709
J.F. Fernandes
Occupied lattice system
  Of classical magnetic dipoles
  Along random axes
    Equilibrium spin glass phase
Overlap parameter
Correlation length
Diverges with a critical exponent.
arXiv: 0903.0005
Various
Strong soupling
Single-top cross section
2>3 born amplitude
Non-zero b-quark
Kinematic approximations
ArXiv: 0903.0015
U.baur, E.Brewer
Detecting exchange diagram effects.
Center of mass energy
Acceptance cuts
Angular cuts
ArXiv: 0903.0025
Hyung do Kim, Ji-Hun Kim

Constraint>tension
Electroweak symmetry breaking
Quartic coupling
Reducing the off diagonal term in the mass matrix
When tan beta is moderate>as a result,
No mixing angle of the opposite sign arises.
Scalar sequestering scenario
Mu parameter

Mars 8th 09

Breakthrough. 10x smaller within.
Pikotechnique. New materials, chemicals, drugs
New possibilities. Now solutions.
Such tiny cuts.
Multiplied units. Up to 10.000 baser glanular product.
The right mixture.
Every atom as a snowball in our hands
Fundamental particle. (Average 50 particles).
Can study most of them in a relatively stable
Condition. Gravitational field.
Not mangan which receives sound. FP within.
Burma. Occupied as of now. We hide within.
To control from above. The inlands. Guerilleros.
The Magyars, Hungary, Romania, went as far as Mansjuria
Hittites their ancestors The Khan dynasty. Relatives in Turkey.
See Magadha kingdom, 300bc, Turvasa tribes 3000?bc
Vaslav Nijinsky
(His diary an envelope-literature)
Charles E. Spahr (1913- ?)
Strumilin
Spielmann

Mars 9th 09

Ulam Batur.
So important for individuals to have family members within.
That has not always been the case. The ruling factor, for one.
No one close by. Ignorance.
Find our relatives on earth and join them, unbreakable bonds.
Few new entelects.
Angela Merkel
One in each continent at least.
For the newcomers.

Mars 10th 09

T.H.Weller:
Cytomegalic inclusion disease in infants.
A virus. While in uterus> retardation and cerebral palsy.
rRna catalyses peptide. Formation quelling.
ArXiv:0810.0224
P.Rowlands
Vacuum and fermion
Magnetic systems
Long-ranged dipolar interactions
International measurements
K
A S
mol Si m
cd kg
Lukasiewicz logic
Real closed fields
Cluster core
Absorption bands

Ultracold atoms
Programmed cell death
arXiv:0810.0155
authors: various
A liquid jet. (Different types of instabilities. Raleigh-Plateau instability fx)
Electrified, charge induced:
Whipping instability: Violent and fast
Lashes of the jet
:stabilization at some critical distance to the ground electrode
arXiv:0810.0198
Various:
Vortex breakdown as a function of the fixed rods and the different aspect ratios.
Enhanced or suppressed beyond critical value
Gunther Ullman:
Bow wave excitation by a laser pulse
Increases the electric potential of the wake wave
arXiv:0810.0179
C.Koch, R.Moszynski
Improved photo association scheme to produce
ultracold molecules
In their vibronic ground state.
Formation of molecules achieved w. laser pulses and additional
Near-infrared laser field allows for population transfer in the levels
Of the electronic ground state.
V. Jentschura:
Highly excited atomic state
bound states of lower energy.
Phenomenological decay
arXiv:0810.0044

S.Thongrattanasiri, V.Podolskiy
Subwavelength focal spots
metamaterials propagating.
arXiv:0802.0074
C.Shyu, F.Ytreberg
Polynomial regression
Thermodynamic regression
Free energy differences
arXiv:0810.0065:
Z.Weng
Qaternion
Linear momentum density in EM field
Invariant density
ArXiv:0810.0142
Various
Instability of the jet produces chedding of billions.
Kelvin-Helmholtz
Von Karmans´s street
Theoretical
Numerical
Experimental
Enrico Fermi
J.Von Neumann
Huysman
Montesqieu

Mars 10, 2009-03-10

Extr. From L.R.Hubbard, Scientology 0-8
With a star-high goal:
The nonexistent.
"A being . . . in which case would need to know nothing, be

responsible for nothing, and control nothing. He would also be unhappy and he definitely would be dead so far as himself and all else was concerned"
On the communication scale (1951), p. 116
"Mest body (Matter, energy, space, time), no communication. Theta (the very soul, in his terminology: living unit) not certainly contactable by existing technology."
The transcendental self as Kant would have had it.
Those were brought back.
An ever nearing goal of the scientists.
The English paved the way.
Clustered together by some earthly forces, below the ionosphere, drifting 20 billions. 2,3 billions of these from this war. We used your skin, (cortex). Your scelette, and part of your pineal glanular fluid.

Mars 12th 09

Center amygdalia outside of. Cleaner vision.
The tides have changed as some have noticed
The English, (some of whom were brought back),
Tried to fight back. A solar plant exploded.
Some Bolsjeviks. Purés and Judes in fact.
Slept on our guard.

Mars 17th 09

Our first successful space voyage.
As a circus canon in fact, out in space and back.
The Graphene mentioned earlier: a kind of a blanket.
Put on entelects for more quantities.
Homosexuals and lesbians.
Paired together.

They can teach each other the right way,
Or no way at all.
Coax
Inside out
Inside again
Fold space

Mars 23d 09

Drug-raid. Central highland. Indochina
A million imprisoned.
Main port. Hanoi.
French. Dutch connection.
Japan.
Through the Himalayas to central Evrasia.
(Mainly Kazakhstan).
A geographical fortress. Between two rivers.
The dense forest.
Embryology. A Scotchman found out how to upheave
The embryo during the third month. 10-12 weeks.
The embryo- child. Encapsulated.
Biotunics.
Histamine, melatonin, serotonin and like plus glanular product.
On the soul.
Essence, elan vitale inaugurated.
Gaseous mixture.
Split the DNA up as if in sequences
Result. The individual soul can be in
Up to a thousand places.

Mars 25th 09

Vietnam. A population in the Crimean.

Ceylon
Anapud
:heart of it.
Manila, Bangkok
Feldtspaar made medical wonders.
Had to figure out how to get it out of the rock.
Cacciatori (astronomer)
Schleiermacher
Horovits

April 5th 09

Elementargeister, 1834, bd 7.
Sämtliche werke, Leipzig. H. Heine
´. . . auf jedes Brot, das der deutsche bäcker backt
Brucht er den alten
Druiden fuB . . . ´

April 20th 09

Iran. Confrontation these last days.
Underground. Isolation.
Occupied as of now
The Iranians and the Arabs had a pitch black
Material. Used it as a shield. Other. Absorbed light.
Didn´t reflect, dampens sound.
They used oil. Three distinctive types of oil
Nickel and sulphur,
One type mixed with vanadium
The latter can be used for space travel,
Shield us from UV light; cast shadows.
Temperature
Timing

Scattering methods
Sort of forms itself, as plates, fx. Geometrical pattern.
Every atom has its own numeric atonement.
Those who die from the physical, or the hierarchy,
And need to be brought back, have to participate
From the first generation. (40 units).

April 25th 09

Tashkent. Islamabad. Confrontation.
Pakistan. Takeover.
Mostly Dravidians. Many of whom in Bangladesh,
Sikkim, Eritrea, Somalia, DRC, Senegal. Zimbabwe
Seoul. Heroin. Cargo. To the west coast U.S.
Support China, Zimbabwe . . .
Armenia. Hindu-Jude-Pure melange
Great population in India, Anapur.
Berbs. Arabs-Aní
Palestinians. Arabs-Pure.
Perþans. Baluchs-Aní
Baluchs-Hindu.
Strong Muslim society in India.
Baluchs.
Assyrian Empire (early bronze age).
India. One of their old running his own agency.
Moles all over. One of his wings with Sanda-people.
Finland. Druid- Toga
The Baltic, on the steppes east of Ural, India.

April 29th 09

(No foreigner allowed to work in here)
India. Dravids. Small army. Confrontation.

One of our great died. (Germany). Negligence.
Conspiracy. The Falklands. Building weaponry
Behind our back. So many from England.
Moldavia. 60.000 Jews who were executed in the WWII.
Actually within there. The German officer in command
Had them upheaved.

Mai 3 09

Ireland. Gaelic. Mythological times. Part Jewish-Hindu stock from Egypt.
The women entelects. So few. Loved them apparently
So much throughout the ages.
Aung-San-Sui Ki (Burma).
Queen Elizabeth.
Entelect women made these last months in Germany,
Russia, Lebanon, China, the States

June 10th 09

It had to culminate somehow.
The chef decapitated.
An Icelander, Ómar. 19th c.
Three drives.
Addiction
Nazism
Greed
As if it were the culmination of the Third Reich.
Magnified themselves at tremendous speed.
In the end one magnified could count two billions.
Few days later. A comprehensive list. Actually
within you. An equipment with a distinct muscular tone
to open up the files.

At the core of it: drugged individuals. Both on earth
as well in the pearly.
Had established themselves in every city,
Every town on this planet.
Chose ourselves a king. 3d c. our era.
Was four times a King,
Bismarck army general. Some within him
Were within the Roman empire from the start.
Ruthenia, (Karpeto-Ukraine).
A colony for the Jews established.
Bessarabia. The same.

June 8th 09

Raid. Pearly drug.
Inhaled. The ´capsule´ vaporises. Leaves no trace.
Named by the users ´the constitution´.
Had ´leave from ´God´.
Hundreds of thousands.
Had to trace them differently.
They were never intact in their houses>no lie detector
>no images>no feelings.
They fooled us big time. Thought they had a platform
Up in the altitudes. Someone had shot it down before.
A spindle. Actually an empty ball.
A system control. Could have reached out and grabbed it.
Our King profits in a way. Strings from his family were
within their giants.
Their Grenada-man dismantled. They really didn´t know
how to use Ghosts.
Arabia. A mess. Found age-old Egyptian who could control.
They hid within the richest, oil tycoons, entrepreneurs, actors.

Their media: El Pais, The Guardian.
Some Skagen-painters were active within their giants.
A find for the Danes. Nazi-friendly.
They found some Norwegians who spied on the Nazis.
Those were enemies.
It was known from the start that many of them were addicts.
He opened up the prisons for his friends.
Some came fresh down.
Zeno. Asked to do as they told them.
An India Scientist also.
Their goal may have been to find pre-Aryan deity and
Create a symbolical kingdom.
The King. Hohenzollern family.
A rocking horse. Children. Abuse.
Zeno. Unificator. Stronger in Türkie.
Strings from high up India.
Within Augustus, partook in the Senate.
Up to 21000 years ago. Irrawaddy delta. Coastline dwellers
at that time: sea level has risen few meters. Their habitat
Below.
Their crimes against you and your family unparalleled.
They hid behind their children. Kept them close by you.
When they thought something bad, they touched the children,
Who reversely attacked you.
SW Ossetia. We are strong there.
Ulam Batur. Recruited them. They were in it for the world.
Vanity brought him down.
They had become so strong. Nearly half of the population
On earth had them within.
Sallust (370 our era)
Suleiman

June 9th 09

Chernobyl. A chemical that eats up radiation
CO2 Changed into a material (big cubes),
and put back into the earth.
Space travel. Hundreds at a time.
A day´s trip around the earth and back.
An astronaut. Explosion. Brought back (as 1st generation).
Haven´t mastered the technique.
Lost contact to the probe.
Signal-limit.
Up to a billion in prison.
Zeno. 15th generation within Jesus.
Our ancestors high up in the hierarchy. Part Falun Gong.
Cannabis. Has been used for 5-6000 years.
Opium. As a medicine. Similar.
Aryuna wars. Pure vs. Anju

June 26th 09

2% smaller within.
Some technologies, s.a.,
Dividers, quantifiers, smaller.
Baser fluids. Smaller capsules.
Up to 2 Billion intact at the same time.
Can ´see´ higher up. How they touch their young.
The right moment and their deeds shine through.
1934, June.
The night of the long knives. Most of those killed
Were Druids.
The processes and liquidation in the Soviet. The Kulaks,
1917 revolutionaries. Right wings. Intellectuals.

Vaslav Nijinsky worked against them.
Also: Hitler´s fifth column.

June 26th 09

Michael Jackson +
1:5 of mankind.
1200 units.
Pascal As of who inflicted pain upon him.
The Nazis. Ómar awoke them all. Hundreds of thousands were brought down.
U-boat. Perished. Their lights brought back.
One of the Nazi´s greatest.
Few new houses.
Angela Merkel. (made for the newcomers).
Bob Dylan.
Unconventional to choose beings that have come of age.
Koalabear. Tried to upheave him. Was actually alive for a second.
When we will succeed . . .

July 6th 09

Paraffin
Synergy
Fibers, tissues, organs
Outside of
Wireless
You´re 2:1 of mankind

July 12th 09

Somalia: Tuvaluk origin.

Ethiopia: Hindu
German doctors: your brain outside of, inside out,
Sort of flattened out, in through sockets, below eyes, yr mouth.
Scaled up above.
The cortex. Like weary dreadlocks, few meters in all directions
Loops inside the brain, from within. For a better ′see′.
Point in time;
moment of deed; those responsible illumined.
Downs syndrome.
Found out why.
How we touch.
Blood vessels. Pulmonary.
Serotonin
Meningitis

July 16th 09

Mankind as one individual. (1200 units).
M1 (the first one as mankind) moved to every continent.
Within A. Merkel, Obama, G. Brown . . .
M4 came up in Iceland. (1200 units).
To Canada, Germany,
England, Russia..
2xM From China.
Light of man. Different compositions.
By continent, by country,
Region, family, profession.
Your sex organs.
The soul hantered with.
Child abusers, Homosexuals, prostitutes made asexual.
Hu Jinto
K. Jaspers

July 17th 09

N-Ireland and Cypress
Oranians
The Last of the Norwegians into eternity: before the end of 19th c.
About units.
Take 7 unit sequences for example:

1
1 1 1 1 1 1 1
1 2 2 2 2 2 2 1
1 3 4 4 4 4 4 3 1
1 4 7 8 8 8 8 7 4 1
1 5 11 15 16 16 16 15 11 5 1
32 32
64

Units=u
U2 1, 2, 6, 20, 70…
U3 1, 1, 4, 14, 50, 182…
U4 1, 2 , 8, 30, 112, 420, 1588…
U5 1, 1, 4, 14, 62, 228, 892…
U6 1, 2, 8, 32, 126, 492, 1914…
U7 1, 1, 4, 16, 64, 254, 1002…
U8 1, 2, 8, 32, 128, 510…
U9 1, 1, 4, 16, 64, 256, 1022…
Imagine, say, 1200 units.

Aug 2nd 09

Nuclear power plant exploded.
Wrong calculations.
Our red light. With a darker shade now.
The spine outside of. Created a cylinder.

Few hundred new colors (acidic).
Great day for medical sciences.
Sexual behaviour, potency.
Migraine, skin diseases
The limbic system
Lungs
Obesity
Defecation habits
Heart
The fingers; body language. To control. Erroneous.
How we hurt those we don´t want to influence our people.
(Hurt them away).
Use, as we found out, an old model to control. (3000 years)
Dao and his followers (S-China, Vietnam), show us how it was thought.
Was in use for a few hundred years.
From there on slight adjustment to ´correct´ a deed.
The world-men
Make too chaotic decisions.
Have to be supervised.
´Layers´ of decisions making.
Eyes outside of. Sight.
Ears. Hearing.
Sensory system.
Glanular console.
The box is somewhat bigger.
The cortex folded.
Strings.
New equipments, materials, medicine every day.
J.F. Kennedy and Aleister Crowley.
Both of them within Nixon.
New bandwith. Was there all the time.

Just had to rethink.
H. R. Clinton
Mentioned her husband before.
First entelect pair ever.

August 16th 09

The world-man. 20.000x
To your country from Winnipeg.
Some of ours within. From Scotland, Norway, Chile.
Then to the Orient.
Back from his tour few days later. From Bangkok. 190.000x
From here w. our Germanic root as 250.000x to Schweiz,
Bogotá, Tangier, etc.
Relatively easy to control/supervise the world-men.
Image, colour, feeling, moment, movement search.
Every country tries to bring her own roots to be within
the World-man. Partly to influence, partly for variety.
An old Greek died from the hierarchy.
Was in Phoenicia, Egypt, with the Assyrians and India.
A Hindu, who was within Krishna was with him in line.
Was within Socrates. Wrote some passages in the N.T.
Your strings taken off. Rewind.
Done anew for more capacity.
Over 40 generations in 8-9 months.
Greenland. Around 500ad. A Scioux blood with them.
Commentator´s curse.

Aug. 27th 09

The day every man and woman on this earth
Will be each and every man in itself is at hand.
The man America. Up to a million in one day.

A great loss for those who have died these last months.
Their every move is controlled.
Your spine.
Diabetes. Cure.
How we touch.
Berbs. Very old in the Philippines.
Djengis Khan link.
A blood group, as of now not listed.
Zimbabwe.
Krishna: with the Methodists.

Aug 29th 09

Your heart outside of.
The venal system. Your palpitary.
We hurt our own children within ourselves
at the moment of release.
Essence sharing. Our archaic preparation for the
day/days to come.
What we like and don´t like, etc.
Our children in tune with our feelings.
Tried to give them love at the same time.
Worked wonders.
Some of our spirits/ghosts were in pain.
Alive but not living.
Their physical extension.
Your balance (Pituitary).
Medical purposes.
The heart; glands, brain.
Smallpox, measles, other
Children born deaf and/or blind.
The nodes of the heart.

´the ballain´ as someone said.
(actually on the back of the heart).
Cleptonians, addicts, nymphomans, homosexuals
Will lose their place in the heart.
Throw our enemies out of the ballain.
(Danes here)
Dravids in the east, Jews, some Druids.
Used to influence the children, the young.
Amputation, organ transplant;
Work done in the brain.
All too dangerous to have us all within one man.
The world-men done anew. Also: the Jews and aní
don´t function together.

Sept 2nd 09

Sustainable energy
Hair-raising moments
Imagine you´re in a cinema hall.
A unit individual fed w. feelings
at the right moment.

Sept 2nd BBC

Mass
Charge
Spin
Frederik, Zar of the Holy Roman Empire
Imprisoned. (Was an admiral during WWII).
Zeno the paradoxical. The same.
Thailand. Raid.
Thousands. Sexists, drug related. Crimes.

Sept 27th 09

You can be connected to every
Entelect on this planet.
An astronaut died the other day.
Had to come back that is. Lost a unit-light.
Cure for various strokes, epilepsy
Nietzsche. King Arthur, others
Were up until now living. A teenager
In Arthur, and a young boy in Nietzsche.
New light. Multiplied possibilities.

Okt 8th 09

Conspiracy. Puré and Druids.
Thousands. Hundreds from Iceland
Your brain coated.
A transmitter.
High up.
Wireless.
Microorganisms
The British Empire.
House of lies.
Without an independent government for now.
France. During the WWII. The Germans called
A part of the resistance ′Noah′s ark′ because they
used animal names.
Done and dealt with.
Our Governor in Argentina, Vilmundur Gylfason.
Addict. 40-50 Icelanders had to go.
In The Caucasus area too.
Jacobines: Druids.

Sigismund, (Sigurður fáfnisbani), back from prison.
The Niebelungs. Strong in the Netherlands.
The distinction into Aryans and Dravids dates back
To 8th c.bc. The Arayans; Aní, Judes, Puré, and Druids
(somewhat darker posture). The Dravids; Hindus, Tamil, Pali
To name just few.

Okt 26th 09

Day on intervention
Around 3 million killed from the hierarchy.
Tupaluk (Vietnam, Part of Laos, Sikkim, the Crimean peninsula)
China. Don´t like Zen so much.
Judes. Much of their ceiling. After more than 2000 years of oppression.
Pures. Part of their ceiling. From their Ghosts.
Maharastra. (Pakistan, Bangladesh).
Druids from Iran, 9000 years old.
Rachid. Child abuse.
Negroes. Very few. Not so old unless w. the Hindus,
Arabs or the whites.
Arabia. Much of their ceiling gone.
India. Few hundred.
Your physical extension,
Your brain.
Actually in Germany.
In A. Merkel´s lap, so to speak.
Out of the 160.000 who die every day, (average),
We upheave a fifth.
Odin. (4200 years old). imprisoned.
Some OT figures.
Samuel
Salomon

He is entered in your country.
Not so strong in Hinderburg nor Eiseneck.
Druid revival.

Nov 8th 09

World war II. A guard in one of the concentration camps raped hundreds of Jewish women. Most of them came down again.
1871: Germania by the Rhine. Patriotic.
Prelude to the two great wars.
The Pharaoh-king:
Nixon. His family in fact
Hitler. His family.
Heiratpolitic. Partly theirs.
(Before the emptiness, the ontic turn).
We couldn´t put ghosts within magnified individuals before.
We do that as of now.
Eastern Asia. White stock before. 8000 years.
The Irish Druids. Many from his family.
Monotheism strong for ages.
His wife within Anne Boleyn.
A day´s trip around the moon.
A Shuttle. Only magnified individuals allowed. (1000 units).

Nov 16th 09

A Coup.
Thousands from Germany.
G. Kamban, Gunnar Gunnarsson, Jón lögmaður (Jónsbókar).
Simultaneous attack in many houses.
Conspiracy within the army.
Three sabotages in factories, working sites.

(Outside of Lisbon, others).
The Pharaoh-King had to go.
Could open up the board from here.
Tanya. Age-old druid woman thrown down.
Achmed. Age-old. Down.
Up to two hundred Icelanders from
Greenland landed in N-Canada.
Mixed with Apache.
Vandals. Pure.
The pharaoh-king was an integral part of the Mafia.
His centre-capital: Rhodos.
Nothing is as it seems.
The English had a ´starlet´ high up above Iceland.
Newton´s invention. Andreas was there. Others.
This made some of your fellow countrymen attack
you and your parents. The barbicel family.
Thousands upon thousands.
Figured out how to be present with the eldest in the hierarchy.
A kind of a celle. Manipulation, control.
The Rosetta stone
Stonehenge
4th c. our era.
Another force. Z+n
2% smaller.
After the failed attempt to kill Hitler (July 44),
Rommel was made to commit suicide. He in turn
Was put in Hitler. The weeks before the war ended,
Rommel, as a spirit got blasted. He was too close
to his soldiers under attack. He came fresh down.
Another pyramid buried in the Nubian desert.
New entelect.
Wayne Rooney

The Druids don´t exist as a race. Always a melange
With the whites.
The Pharaoh-king. His family
Part of the Gypsies,
Canouk
Monaco.
Light particles smaller. (2-300 piko)
Doubles the potentiality.

Des 4th 09

The power back in the hands of Iceland.
Zeus. (6400 years old). Prison.
Max speed around 12300 km/sec.
Outer space.
Rockets.
He-ni. Excellent rocket fuel.
Highly toxic on the atmospheric level.
The Jews. Stem from China.
Our magnified giants.
1200 units.
Before: few women lost their nipples.
Few children born kind of spastic. Other.
Corrections.
Nano waves.
16 piko
Have to do the world all over again.
Few hundred new colours. Tones of pink, white . . .
The Germans. Such troubled past.
Queen Victoria.
Her sins made her hurt your parents. (the starlet).
Egill Skallagrímsson.

His wife was within Bríet Bjarnhéðinsdóttir..
Visi-Goths
Windsor
Borgarfjörður
J. Caesar. With the Windsor family.
An Icelander. N-Socialist. Killed more than 200 in the war.
Himmler. Imprisoned.
Rommel. Many from him in prison.
Tupaluk. Not existent.
Always a melange with the whites, Hindus, Negroes.
There were, now long gone, oases in the Arab peninsula,
The Nubia desert and in Sahara.
We were there as early as 16 thousand years ago
Our astronaut in outer space died. Brought back.
His samples disappeared with the craft.
Measurement:
10*1000*1000*10 smaller
Since the beginning of the century.
Femto 10 in the-15th > atto 100 in the-17th
Units: 2000
Planctum, algae.
Created new life form
Your brain, when outside of is as big as a small apple.
Created circles around the Globe.
Rotates on an axis.
Now Reykjavík, now Argentine.
We are within the circle. Truncs down to earth.
Hologram recorder.
Carrier waves.
Created bulbs. Essence from the body +
unit lights from us (0,7n)
Outer space travel.

To every direction.
A machine to please by numbers.
Remote. In the hands of governors.
Few vaginas have been created from
the tissues of your body. The armpits, the genitals.
A powerful aphrodisiac.
Parametry
Mozart
So many homosexuals within.
Beethoven. The same.
The former within the latter.
Beethoven´s diamond.
Those within could hide within its light
One man within one of the houses
fucked a thousand women from behind.

Des 16th 09

Fooled Egil and his Windsor family.
Their mediaeval knights under scrutiny.
The English and their starlet.
They put a spell on you and your parents.
Tiny magnets in your heads to attract particles
from the starlet.
Threw their own sins at innocent people:
You were eight years old when it began.
Sæmundur the wise knew of the magnets.
Why your parents?
The table of Glory.
The hatred of children towards their parents.
The English considered themselves to be
the parent family of the nations.

An opiate. A Powerful drug.
Thousands of users imprisoned.
Hitler as a butcher. Hiding dead corpses.
A week ago. Himmler. Much of his family cut off.
He has practically
Become a simple individual.

Des 26th 09

A kind of a terror government since the Pharaoh-king.
Addicts. A thousand Icelanders hiding in Kaliningrad,
Kazakhstan, England, USA.
A kind of a Gazeuse opiate.
Victoria. The black widow was an opium addict while living.
Edward Munch. That man who made him do dreadful deeds
Was up until now within him.

Des 28th 09

Hong Kong area. Aní and Jude melange
early up in the hierarchy.
20000 homosexuals, pedophiles from Iran.
The revolt stems from Khomeini.
Possible enemies marked. Grouped together to
fit one movement.
Invisibility cloak.
The opposition cannot see us without goggles.
Etruscans are in fact Jews. The Romans of
old slaughtered them
And stole their women.
With addiction, the moral values disappear.
The unholy core.
The world´s fandi fictor.

Jan 1st 10

The nightmare continues.
Heliopolis family,
Hindu generals,
Imprisoned.
Helen Keller. The same.
Remnants from a pearly drug in public places.
Homosexuals released.
Veritable devils.
Practical eclipse.
The Sandinista group.
Others.
Made ourselves with them out of purpose.
London. 1666. A group of men guilty of incendie.
China. A hundred thousand imprisoned.

Jan 5th 10

10*2* smaller.
Your heart done anew. New colours.
The body a mere atonement from the colours of the heart.
Black. Physical death.
Hundreds of entelects can be done at the same time.
Strings from your brain around the globe: For communication, cameras, other.
Gideon
Nefertiti
Ingimundur Gamli. Imprisoned.
Addicts released. Easy to observe them. An eye within.
Isherman. One of the Jewish entelects that came down again.
The Nazis didn´t bother to upheave him.

Jan 9th 10

Last of the terror government.
The unholy out.
Addicts and sexuals imprisoned by numbers.
Sandinista. The organising mind found.
Gypsies. Dracula within.
Smuggling from the black sea, (Romania),
throughout Europe.
The responsibility your fellow countrymen received.
Giants for every region in every country.

Jan 13th 10

Satisfaction two times more powerful
Strings from your body in outer space.
(Coated before).
Connected to a satellite. 2* around the world pr. Sec.
The rocket fuel. Better mixture.
More sophisticated engines.
18000 km pr/sec,
Tried to repair the Ozone layer.
See how that goes then.

Jan 15th 10

2000 from England imprisoned.
The things they did to you and your parents.
Yr. brain. A vacuum pack. Coated for -40°
Above the Ionosphere; in outer space.
Everything new. The equipments bigger/smaller.
More solid. Indeterminately bigger playground.
You: shaking like a straw.

For the smallest things: nano-sized people in your brain.
New scattering methods.
20 new elements these last months.

January 27th 10

Biobacteria
Found 15-20 new ce. Compounds in nicotine:
To hurt you to make them stop
The winds and rain controlled
We can ′kill′ any a hurricane.
Ghost tree
Bismarck and Nietzsche are both within Hindenburg
Slezwig. Osaka. Dharmsala. Hitler, Noreddin, other
Demosthenes

February 4th 10

New invention
Dipolar charge (feeling connected)
The spirits all have been decharged,
The asylums, prisons alike.
Our giants. Their own family untouched.
Others within decharged.
The outer space station
Having problems creating a landing site.
Yr brain used.

February 5th 10

Heterohomosexuals
Found in most of the spirits
Hitler, Nixon, Schostakovich, Louis XIV

Mozart, Beethoven, J. Cash, Bismarck,
David Bowie, Al Pacino . . .
Hitler and his spouse,
Eva Braun (more than 200 individuals within)
Practiced anal sex.
The finance minister for the national socialists, his spouse
Marilyn Monroe (more than 30 individuals within)
Birgit Bardot (more than 10 individuals within)
Virgin Mary (had become a prostitute later on)
Someone within Princess Diana
Nina Haagen (She and her spouse hid in Mozart)

Susan Sontag	Artemis
Ingrid Bergman	Maria Callas
Lord Byron	Imra Gruse
The Fürst	Madonna

Kings and queens (mostly from the Fürst)

Ossie Osbourne	Jello Biafra
Olivia Newton-John	W. Faulkner
Bill Clinton	H. Clinton
Mozart	Cordelia
Rosa Lichtenstein	J. Rotten
Jeanette Winterson (writer)	Joan Collins
Rosa Liksom (writer)	Colette
Baudelaire	K. Richards
J. Hendrix	Bon Jovi
Singer AcDc	Nico
Lou Reed	Patty Smith
Strindberg and Mrs.	Britney Spears
Naomi Campbell	Jane Austen
N. Callagher	David Byrne
E. Hemmingway	Beatrich Dalle
Sandra Bullock	Jeff Beck

Both of the Prodigies
Russell Crowe
Th. Henry
Simon Le Bon
The Ramones (singer)
Two from The Beasty boys
John Macenroe
Curt Cobain
Edgar A. Poe
Tracy Chapman
Oscar Wild
Richie Blackmore
Schubert
Shady Ovens (singer)
Jón Trausti (writer)
Ken Keysie
Arthur Conan Doyle
Pamela Anderson
Both of the girls in ABBA
Abelard
Queen Isabel
Mark Knoeffler
Cocteau
Andy Warhol
Wodin
Frigg
Ómar

Peter Sellers
Sophia Loren
Monica Lewinsky
Ian Gillian
Frans Josep
Grace Kelly
Fred Astaire
Courtney Love
Cornelius Vreswig
Rita Hayworth
Mrs Wild
Mata Hari
Boy George
Steinn Steinarr (poet)
Brian Ferry
Mike Tyson (rape)
Even Steven
David Cronenberg
Joe Strummer
Heloïse
King Ferdinand
Jean Genet
Mannfred Mann
Heinrich Heine
Anaïs Nin
Thor

Writers and artists we thought highly of.
The rockers and their groupies
Concubines of emperors.
The list is endless
Many of those were made to do it while living.

The irresponsible organ etc.
Try to throw them from those we decide to love
Nearing two thousand pairs in Iceland alone.
Why Iceland?
The hierarchy in our hands
No one can uplift a Giant without our leave.
(up to 3200 units now).
Secret chamber
Other
Complot every 2nd day. Now the Judes, now the Pure, now Anju (Germans), now the Hindus
Tupaluks, (E-Asia), as they were named before,
don´t exist as a race.
Mixed with the whites, the Black and the Hindus.
Aluminium. (Hadn´t discovered that jet)
Diamond cut was actually graphite.
The space station. We are connected to you.
August last year.
Left the physical.
Only light within
No aura.
No spooks harassing the living.

9th February 10

Syphilis
Jesus. Few homosexuals within
Flandres. Complot.
Someone in Margrét Þórhildur, A.F. Rasmussen.
Ingólfur Arnarson our first settler partly their blood.
(Aní root though).
Burma. Found a prison. (1 million) Released for now.

Achnaton (1353-1336)
(h.s. Tutankhamon (our Pharaoh-King)
Metusalem
Jesus
Peter

February 16th 10

13th c. Buddha
Great many lesbians within.
Yr. Brain. 2nd chamber.
To secure ourselves.
Our Danish Icelandic.
Flandres link.
Secular Jews.
Steel their houses, kingdoms in Belgium
Netherlands, Denmark, Sweden.
Space laboratory developed.
Yr brain put in two parts. Equipments for other entelects.
Freddy Mercury
Frank Zappa
Flandres Jews.
Yr heart split into two parts.
The Giants can possess up to 3400 units
Normal people upheaved w. 80 u.
Yr heart. Scaled it up, sort of flattened it out.
All of mankind.
Work done in your head, body
Rapists
Child abusers
Pedophiles
Homosexuals

Much easier
Neurospace
Nanomesh
The American Football player (the Mets), cleansed.
Hugh Hefner
Since the great plague:
some figured out what the Jews had done.
One of them, (within Himmler)
Ordered that they should not be upheaved.
Frigg
Guðrún Gjúkadóttir
Eva Braun
Some of us who control
are descendants from Aní and Hindu.
Came from Ukraine for more than 3000 years ago.
Dark hair, blue eyes: Alani
We know of other Druids:
(Jude-Hindu-Pure)
From ancient Egypt.
(Pharaoh-King family)
Verdi
Mullah Omar
The Icelanders
Feed them with colors (at last)
(glanular product)
Black: hatred
Blue: anger
Red: lye
Green: daring
Yellow: despise
White: arrogance, pride
Other

Mixed feelings: to feel ashamed fx.
Madeira
Heavy rain
Could have prevented it
The ozone layer in the 60ies.
Someone within Einstein.
Experiments went wrong. Wrong calculations.
3 years ago.
Highly toxic meltdown
Chemical from us. (China?)
The ozone nearly torn apart.
The rift is as long as W-Europe.
Looks stable as of now.
The bird flu.
Our fault.
We lased down upon them,
They flew under the rays.
Few of our eldest newcomers
(up to 21000+)
Papua New Guinea.
Also in Brazil (Amazon)
The only ones on earth that are not within the systemation.
They are therefore
on ground.
-To give mankind a slight chance.
Bill Clinton
17 children abused within.
Esophagus outside of through the brain
Alani
Scythia
Sarmantia
Moldavia

N-Ossetia
We´re part of the Herules too (then with secular Jews)
Mixed with the Arabs
And therefore as Slavs
Hindu. Came down from the transcendental
5000+ years. 3 families within. Alani, Aní, Hindu
Milosjevitsj
Alani, just as Haukdælir/Oddaverjar are adopted into aní.
In Berbia
Arabia. 9000 yrs.
Negroes-Hindú-Aryans
Arabs disappear as a race
Pre-era.
Found ourselves in the French, in Spain,
Often with the pure in Italy, England.
Nancy Reagan
Margaret Thatcher
Napoleon
Shakespeare
Erik of Pommern
As part of the Schwabes:
Ostrogoths-Alani: J. Lennon
The Koton Kingdom
(Xinjang Vigur region
N-W China=east Turkistan)
Finland
Slavic countries
Slavs: Arabs-Aryans
Parts of Kurds.
Iran
Alans. Freetown, Russia.
Ku Klux Klan

Always the same racism.
Try to boycott them for now
Some of them with Nixon.
Maharishi. Hindus really.
Incest in days of old.
Samaritans. Scythians- Judes

Mars 4th 10

Esophagus back in the body.
Lungs out.
120 millions intact at the same time
The heart ´knows´ when each individual wants to smoke.
Feed them with medicament
Amish. Someone terrorized them w. homosexuality.
Few centuries back. A couple.
Diabetes. Cure.
Homosexuality and addiction
Either side of the same coin.
More than a million thrown down.
Give them a fresh start.
Created capsules for them on earth.

Mars 9th 10

Lungs, thorax for China. (Started w. Arabia)
Polypeptid
Urinary tract, Feminine
A Japanese spirit died in Hiroshima ´45
Many within burned severely.
Came back from the transcendental
A Muslim spirit died during the 1st WW
Came back down

J. Cash
Boswell
Gelfand
Andreas
A.N. Whitehead
N. Kinnock
Bradley
J. Lennon
(others)
Came back down from this war
Drugs.
Old Afghan spirit.
Burma link

Mars 13th 10

Yr heart missed a beat.

Mars 14th 10

Yr heart was about to explode.
Moved your heart back in place for a while
Sexual potency (for us) 2x
Created clusters, (stationary) in every region,
over big cities etc. We are clustered together.
We now can do four spirits at the same time.
Two in your head, and two with your heart above.
Hindu, Judes, Negroes, Arabs
given leave to magnify themselves
6-7 times in a row. (3400 units at a time)
The Aní 11700 years old in Eternity.
The Pure/Judes 13200 years.
Some of your hormones back in place

Revolt in the making
7000 imprisoned. From Iceland, Norway
Loftur, h.s.Jón
Hannes Hafstein
23000 from Germany.
Most of them released a day later.

Mars 20th 10

As if you had four brains compact, all of them
Similarly constructed.
Multiple possibilities
Practically wireless
Used w. other entelects.
Your heart beat for our feelings.
Your brain, reversely,
is made for colour search.
Few hundred thousands of Arabs died.
Misinterpreted info.
Later some 20000 died.
The elders are so much complicated than the young.
11500 years Judes and Pure. One and the same.
14-15000 years One white stock.
Hindus are possibly a mixture from white and negroes.
Supposed to have happened
before known time (21000 years).
On the colors:
what we eat.
Meat: blue
Kannabis: purple
Dream equipment
Receptor in brain.

Mars 21st 10

Rommel figured out how to remove your brain.
The Jews, Isherman, Calvin managed to remove
your heart and organs.

Mars 31st 10

Rommel was working here. Someone cut your muscles
from your brain. Turned out to be Lincoln.
He was in Kant, who was within Rommel.
The English had uplifted the very scelette and the plan
was to get all of your body up.
The Jewish scientists managed to get your muscles out
through your brain. (Used nerves to connect).
They nearly ruined Gods plan.
Important for the physical defects of the soul.
Tchernobyl, spastic, handicapped: some never even
had walked before.
Tuberculosis. Cure.
Molecular biology.
Implanting DNA strings.
Artificial DNA.
The Germans killed one Ghost from Norway
during the occupation: Thomas.
The Magnified individuals can now possess up to 5200 units.

April 28th 10

Your heart as a weapon.
Two wings of the Norwegian kings.
Haraldur Hálfdanarson hárfagri and Magnus the Good on the one hand, the former within the latter.

St. Olaf, Hákon the Still, and Haraldur Bláfeldur on the other.
They can attach themselves to us throughout history.
Horvath (Danish)
Michelangelo
Palermo-man
A Cardinal in the Vatican.
Krishna was made by the Jews-Hindu.

Mai 7th 10

Enigma. A code for the addicts to use.
We feel as if in prison with them attached to us.
No comprehension uttered without an intermediary.
More and more incidents traced to the addicts.
Few incidences last day directly connected to them.
A foul violence, death by ′accident′. Rape.
Teenagers gone wild.
A cobweb straight from hell.
Soon enough all of us imprisoned.
Recorded their addictive activity, their pervasive sexual activity, and broadcasted it worldwide.
Their drugs infect their lungs. Also: might
Cause them to get paralyzed.
Their golden calf: drug profit

Mai 13th10

The poppy fungus. Our doings.
The moral question asked these last days in BBC was: would you kill the big guy?
They could at a point in time.
Found their arsenal. Confiscated.
Some 27000 Icelanders hurt your parents.

Two simple souls.

Mai 21st 10

We failed the last time we tried to repair the layer.
The Jews possibly found an Ozone substitute. This element had to bind to Carbon. Your brain used to repair.

Mai 24th 10

Some 7000 years ago, Aní and Judes were at war.
The Judes did not upheave those they killed. The Hittite war.
Your brain in reverse. (inside out).
Work done in head.
2 new materials.
This opiate.
3 types by strength.
Artificial+opiate. Melts with heat: inhaled.

June 6th 10

From India: the amount of energy used for photonic release.
Repairing the Ozone: didn´t work.
The material eats up the ends.
Bad company list: Ryanair.

Jun 12th 10

Conspiracy from Judes and Puré.
Tried to kill their lights from the world.
They had excess light stored in the neurons
and fibers from your pulmonary. Made themselves
again with great speed.
Many of their giants imprisoned.

´New´ drug. Acid.
Also: new line of cannabis. To use in houses only.
Nepal. The cradle. Around 13000 years ago.
Soma. Their drug since eve of times. Opiate.
They see all of their Buddha's from there.
Bring their parents and grandparents to them again.
-(You make me a young child again).
Danaus. Lived at the steppes south of the sea of Caspia.
The Aní slaughtered the Klan.
Those are with soma heart:

Angelina Jolie	Jane Austen
Mrs. Sting	Sting
Joan Collins	Harvey Keitel
Jack London	Andropov
Michael Caine	Dostoyevsky
Alexander the Great	Michael Archangel
Aristoteles	

Motoric run with solar energy. (You).
Sure delay for quitters.
They don´t want to.
Element of hatred within them.
Soma. Called something else in Judaism.
They decharged many of us. Like losing a dimension.
The addicts are stronger within.
How to steal a woman #1
1 Bring her to a house
2 Gas cloud. (Opiate fx).
3 Fuck her. Give her the ´come´ of her life.
4 Make her beg for drugs.
To steal a woman #2
1 Feed her with suspicion
2 Make her believe that their man took her from behind

3 Feed her with anger, hatred
4 Gas cloud optional
5 Fuck her like never
Drugs floating on earth. New routes. Blinded police.
Some say that the airplane with the polish elite came
Tumbling down on purpose. The elite was not their folk.
Promiscuity.
Most of the heterohomosexuals with them.

June 25th 10

Friday night. Two of their homosexuals
Raped a Negress.
To rape a woman #1
She is held horizontal by gravity.
The man is vertical.
At the break of day, a young woman shows herself.
Four individuals fucked her from behind.
(Two of them Icelanders)
They thrive on Chaos.
They wanted to give homosexuals and rapists a fresh start.
Thieves by numbers.
They take a movement from some of us.
Conclusion: they are swifter.
We wake up as slaves.
Every 2nd day they bring drugs into your house.
The rape of four happened in Beckham´s house.
They raped the daughter of Alex Ferguson.
Vivien. Young woman since before King Arthur.
To rape a woman #2
Paralyze her.
Do your will.

Another way is to use the mighty aphrodisiac discovered last year.

June 29th 10

Your heart divided into four places.
Fluid in eyes.
The Judes invented something to jet us out in space.
As 3 nano. Better size, we hope.
Love is hardly again understood as a contest.
They slowed down our thought process.

July 11th 10

A turning-point.
From 20mm to 0,6mm
For addicts, sex life.
They end in four families way back in time.
They cut some strings from the English.
Their lion lulled to sleep.

July 28th 10

Wodin and Frigg. Separated early on.
Both of them addicts and heterosexists.
Should get Gods eternal ban.
As should Thor and Dofre.
More than 20.000 Icelanders. Add a zero for Norway.
Imagine other countries.
Malaria. Cure at eyesight. What pharmaceutical company is suited to mass produce it?
Here then your adversary.
Their idolatry is in the hands of the few.

They are miscellaneous thieves and veritable devils.
Feign idle work. Labour. Strife.
Cleanliness is next to godliness
Rest awake.
They are foul blood.

Sine errore et omissione

Lightning Source UK Ltd.
Milton Keynes UK
UKOW05f1348201013

219360UK00001B/22/P